50 YEARS IN SPACE

WHAT WE THOUGHT THEN... WHAT WE KNOW NOW

Dedicated to the crew of the Space Shuttle Columbia,
who died during re-entry on 1 February 2003.
They accepted the challenge and the risks,
and their wish was for their dreams, and ours,
to become reality.

50 YEARS IN SPACE

WHAT WE THOUGHT THEN...
WHAT WE KNOW NOW

DAVID A. HARDY & PATRICK MOORE

AAPPL

Published by **AAPPL**
Artists' and Photographers' Press Ltd
Church Farm House, Wisley, Surrey, GU23 6QL
info@aappl.com www.aappl.com

UK Trade, wholesale and export enquiries to:
Turnaround Publisher Services Ltd
Unit 3 Olympia Trading Estate
Coburg Rd., London N22 6TZ
Tel: 020 8829 3000 Fax: 020 8881 5088
orders@turnaround-uk.com
Distribution USA: Sterling Publishing Inc
sales@sterlingpub.com
Distribution Australia and New Zealand: Peribo Pty
peribomec@bigpond.com

A catalogue record for this book is available from
the British Library.

ISBN: 1904332609 9781904332602

Editor: Paul Barnett
Art Director and Designer: Malcolm Couch
Printed in China by Imago Publishing
info@imago.co.uk

For further information on books from AAPPL
visit www.aappl.com
To buy or license paintings or prints by
David Hardy please visit www.astroart.org

Previous page, left:
Several methods for starflight were explored in the 1972 book, but one that made it through to both editions was this one, of a photon drive. Effectively, this emits a beam of light from the giant parabolic reflector (here cooling upon arrival), which propels the ship – rather in the way that the 'primitive' ion beam motor does for the lunar probe on pages 14-15. The background here is the beautiful Trifid Nebula, which like Orion is the birthplace of stars. The lightship has arrived at an Earth-like world, with its moon.

Previous page, right:
The huge red star Antares ('rival to Mars') is over 9,000 times more luminous than the Sun, but has a temperature of 'only' 3065°C, compared with our Sun's 5565°C Even so, a planet would need to be 20,000 million (20 billion) kilometres away in order to have an Earthlike atmosphere and liquid water. This idyllic landscape was deliberately painted to emulate the rather romantic work of the Hudson River School of artists, who painted the opening of America's wild frontiers, such as Yellowstone Park and the Grand Canyon. The nebula around Antares is hidden by the atmosphere and clouds. *(From the private collection of David Egge.)*

CONTENTS

OVERVIEW 1954-2004 6-7

SPACE STATIONS 8-11

THE MOON 12-17

MARS 18-29

VENUS 30-35

MERCURY 36-37

ASTEROIDS AND COMETS 38-45

JUPITER 46-55

SATURN 56-63

URANUS 64-67

NEPTUNE 68-71

PLUTO 72-75

THE MILKY WAY 76-77

PROXIMA CENTAURI 78-79

ANTARES 80-81

ALGOL AND FORMALHAUT 82-85

TAU GRUIS AND BROWN DWARF 86-89

BLACK HOLES 90-91

ETA CARINAE 92-93

NOVAE AND PULSARS 94-97

GLOBULAR CLUSTERS AND DEATH OF THE SUN 98-101

NEBULAE 102-103

JETTING GALAXIES 104-105

COLLIDING GALLAXIES 106-107

INTERSTELLAR TRAVEL AND COMMUNICATION 108-109

SOCIETIES AND ORGANISATIONS 110

INDEX 111-112

ACKNOWLEDGEMENTS 112

THE CHALLENGE OF THE STARS

OVERVIEW: 1954 - 2004

THE 20TH CENTURY was the Age of Challenge. In 1903 came the first manned flight in a heavier-than-air machine; 1926 saw the first liquid-fuelled rocket. There followed the first flights above the main atmosphere of the Earth, and then, in 1961, the pioneer flight of the first spaceman, Yuri Gagarin. By 2000, human beings had reached the Moon and unmanned spacecraft had surveyed every planet in the Solar System except for remote Pluto. All this was a prelude to what may well be the Age of Achievement – the 21st century.

When we – Patrick Moore and David A. Hardy – first discussed the idea of a book to be called *The Challenge of the Stars*, as long ago as 1954, we hoped that we could make forecasts with reasonable accuracy. First, an orbiting space-station and then an expedition to the Moon, establishing a base, small at first but growing steadily. Building upon the lunar experience, humankind would send an expedition to Mars, perhaps setting up a base there too. Jupiter's satellites and the outer Solar System would be the next targets, followed, perhaps, eventually by the stars.

Our schedule seemed logical enough, and the Moon was indeed reached – in 1969, before most people had expected. But thereafter things did not go entirely according to our plan. By the time the first edition of *The Challenge of the Stars* was published, in 1972, humanity was about to go to the Moon for the *last time* it would do so in the 20th century, in Apollo 17. For motives that were political rather than scientific, the USA had chosen to visit the Moon first in a single, expendable vehicle. True, a year later NASA did put Skylab into orbit, and this was in effect a space-station, but it was not intended to be permanent, and it was not developed. Russia (then the USSR) followed with Salyut and Mir, and by the end of the century the International Space Station (ISS) was under construction, but even now there are no firm plans or commitments to send people back to the Moon or set out for Mars.

There seems to be a 'window of opportunity' within which humans can choose whether or not to become a spacefaring race while we still have the knowledge, the means and the resources to do so. Sadly, this window seems to be closing increasingly quickly. We hope that in some small way our book will act as an optimistic reminder of what lies out in space, waiting for our exploration – and even exploitation, as in minerals from the Moon and asteroids, together with the advantages of solar power. Moreover, if we have the foresight and will to make space exploration happen, it will amaze, excite, and enrich the lives of the next generations: your children and your children's children.

Remember the words of that great visionary, Arthur C. Clarke, written in 1968 when the outlook seemed so promising:

The challenge of the great spaces between the worlds is a stupendous one, but, if we fail to meet it, the story of our race will be drawing to a close. Humanity will have turned its back upon the still untrodden heights and will be descending again the long slope that stretches, across a thousand million years of time, down to the shores of the primeval sea.

Having said that, the fifty years between our first visions and today's reality have seen the most amazing discoveries and advances in astronomy as well as in space technology. In 1954 photography had yet to be superseded by electronic devices, and some of the ideas then

In the 1950s the US artist Chesley Bonestell depicted streamlined, V2 like spaceships landing on the Moon, and many science-fiction artists followed his lead. For *The Challenge of the Stars*, Moore and Hardy preferred the designs of UK artist/engineer R.A. Smith, who worked with the British Interplanetary Society (BIS) and illustrated books by Arthur C. Clarke. His lunar lander bears a resemblance to the later Apollo LM, with shock-absorbing landing legs – indeed, Apollo incorporated many features originally proposed by the BIS.

current seem very old-fashioned today. The brilliant, canal-building Martians of Percival Lowell had been consigned to the realm of myth, but Mars was still believed to have vast tracts of vegetation. Venus might have jungles, or oceans of soda-water; Saturn was the only planet known to be ringed, and we knew nothing about the volcanoes of Io, the lava-flows on Venus, the geysers on Triton or the amazing diversity of the moons of the outer planets. Neutron stars, quasars, pulsars and black holes were not only unknown but mainly unsuspected, while estimates of the age of the universe as we know it were little more than guesswork.

Thanks to probes such as the Mariners, the Veneras, the Vikings and the Voyagers, we have learned a great deal since then, and the developments in instrumentation have been truly staggering. In 1954 the world's largest telescope, the 200in (508cm) Hale reflector at Palomar in California, was in a class of its own. Today it is regarded as being of no more than medium size, and we have of course the Hubble Space Telescope, soaring high above the main part of the Earth's atmosphere.

The second edition of this book, *The New Challenge of the Stars*, published in 1978, included many changes. The sky of Mars became orange-pink rather than dark blue, and Titan, Saturn's main satellite, was very different from the earlier version. But, when we came to prepare this new version, *Futures*, we realized that the relatively minor developments between 1954 and 1978 were in no way to be compared with all those that have occurred since 1978. It seems that almost every week brings its quota of new discoveries – and space art, as well as space science, has moved with the times.

Just as the images from space are now often gathered or processed electronically or digitally, many of the new illustrations for this book were produced on an AppleMac. (This does not mean, though, that they are 'computer-generated'. Although this method of working has undoubtedly speeded up the process of illustration, pushing paint around on art board or canvas has mainly been replaced by pushing pixels around on a monitor!)

SPACE-STATIONS

ABOVE:
NASA's 1970 vision of an orbiting space-station, shuttles and the deep-space vehicle powered by the Nerva nuclear engine. The shuttle on the right has delivered a propellant module for the ship bound for Mars.

FACING PAGE:
Slightly modified digitally (to fit the vertical format), a painting produced for the unpublished 1954 precursor to *The Challenge of the Stars*. It shows the R.A. Smith design, modified by Hardy, for a ferry rocket, and his 'tankers', a version of the von Braun space-station, and Clarke's dumbbell interplanetary spaceship. When this was painted, the only photographs of the Earth as a planet were grainy black-and-white images taken by captured V2s.

OVERLEAF:
Space-station and hotel. The next advances in low Earth orbit (LEO) may come through space tourism; indeed, the first tourist, Dennis Tito, visited the ISS in 2001. In this illustration, two 'Supershuttles' bring up and return passengers; in the distance is the International Space Station (ISS) with NASA's Orbital Space Plane (OSP) in attendance. This vehicle – the replacement for today's aged Shuttle – will separate out crew and cargo functions and incorporate a new Crew Escape System for increased safety.

Fictional space-stations were first described many years ago, but only in the mid-20th century were the first serious proposals made. There were so many unknowns. How would the human body react to conditions of weightlessness? Would 'space sickness' be a real problem? And what about the dangers from meteoroids and the Van Allen radiation belts surrounding the Earth?

Wernher von Braun, the German pioneer who had worked on the V2 weapons during the war and had then masterminded the early US space programme, designed a wheel-shaped space-station. The crew would live in the rim; rotating the wheel would produce centripetal force that would act as 'artificial gravity'. There would be ferry ships (the equivalent of today's Space Shuttle) of 'canard' design, with long rear wings and small wings at the front.

For the British Interplanetary Society (BIS), Ralph Smith designed a space-station consisting largely of a huge parabolic reflector, concentrating the Sun's heat to produce power. His ferries were much more similar to the present Shuttle, with delta wings. Smith also envisaged unmanned tanker rockets taking fuel and oxidant into orbit for vehicles travelling further into space.

In both cases, it was assumed the space-station would be put into orbit first, and that vehicles to travel to the Moon or Mars would either be built in orbit (the US plan) or placed in orbit to be refuelled. Arthur C. Clarke designed a nuclear-powered, dumbbell-shaped vehicle with the crew in one sphere and the fuel and motors in the other, a long tube connecting the two. The 'fuel' – water – would shield the crew from dangerous radiation.

It was clear to all that ferry rockets would be essential to supply a permanent space-station, and accordingly the Space Shuttle was planned in the 1970s. The original design was eminently sensible: both the booster and the Shuttle itself were to be manned and reusable. But, as so often happens, cost considerations prevailed over efficient design, and when actually built the Shuttle consisted of the orbiting vehicle, twin solid-fuel reusable boosters, and an external fuel tank which would be jettisoned to burn up in the Earth's atmosphere. The Russian shuttle – Buran – flew only once, on 15 November, 1988, before falling a victim to cost-cutting.

The Russian space-station Mir was launched in 1986 and remained in orbit until 2001; only in the latter part of its career did it experience serious problems. Mir led to the various designs drawn up in the USA. The Freedom station – quite unlike the von Braun wheel – was planned in 1987, but abandoned three years later as being too heavy, too complicated and too expensive.

The next design, Alpha, gave rise to the International Space Station (ISS). Sixteen nations took part, including the USA, Japan and the participants of the European Space Agency. Much of the ISS was in place by 2003, and, despite the loss of two Shuttles – *Challenger* in 1986 and *Columbia* in 2003 – there is no doubt that it is a triumph of 21st-century technology. Even so, it has many critics, who would prefer the money to have been spent on a return to the Moon – or even on a manned Mars mission.

THE MOON

The Moon, our faithful companion in space, is on average less than 400,000km (250,000 miles) away. It has its disadvantages as a world for colonization, mainly that it is virtually devoid of atmosphere, but at least it is accessible. Its gravitational pull is weak; on the Moon you have only one-sixth of your Earth weight. This is not in the least uncomfortable, and Neil Armstrong, the first man to land there, once commented that 'the Moon is a very pleasant place to work in – in some ways better than the Earth, I think!'

Lunar rocks are of the same type as those of the primitive Earth. Since the atmosphere was lost at an early stage, there is no erosion now; the Moon looks the same as it did long before the Earth was ruled by the dinosaurs. The surface is dominated by impact craters, some of which are huge; the largest and deepest basin, the 'South Pole-Aitken Basin', is over 3,000km (2,000 miles) across, with a floor some 10km (6 miles) below the outer terrain. Many of the craters have terraced walls and high central peaks; but in general the slopes are not steep, and in profile a lunar crater is more like a shallow saucer than a mine-shaft. The surface shows no changes now; for the last thousand million years the Moon has seen almost no activity.

Tidal forces acting over the ages have made the Moon's rotation synchronous; i.e., the axial rotation period is the same as the orbital period (27.3 days), so the Moon always keeps the same face turned toward the Earth. Conditions on the Earth-facing and far hemispheres are similar, though the far side lacks the extensive dark plains characteristic of the near side. These plains are still called 'seas', *maria*, even though there has never been any water in them; they were produced by magma flowing out from beneath the crust and flooding the basins.

One thing is certain: the Moon has been sterile throughout its long history. No trace of life has ever appeared there. Life has had profound effects on the development of the Earth, at least on the surface regions; the Moon, obviously, shows none of these effects.

Below:
In 1970, NASA's 'Plan for Space' included a basic space-station module which could also be used as a Moon-base. Such modules would be placed on the lunar surface by the propulsion module of a 'space-tug', seen landing on the right. The base would accommodate a crew of twelve. In the foreground is a lunar rover, similar to the ones used by the Apollo astronauts.

Slight surface tremors occur, but powerful moonquakes belong to the remote past; there is no danger of one demolishing a lunar base. Neither do falling meteorites appear to represent a major hazard. On the other hand, the lunar surface is exposed to all the radiations coming from space, and in this respect the situation must always be carefully monitored. Obviously, the greatest care must be taken to guard against leaks from the base; everything will have to be airtight. The great temperature range between day and night will govern the choice of materials making up the base.

Astronomers on the Moon will have great advantages. Telescopes will not have to peer through layers of unsteady atmosphere. The low gravity means it will be easier to build large pieces of equipment, and this will be particularly useful for radio telescopes. The far side of the Moon will be the perfect site for a radio observatory, because it is completely 'radio quiet'. It may even be possible to press a crater into service, in the same way as for the Arecibo radio telescope in Puerto Rico.

Right:
In 1972 we chose a site near the Moon's north pole for our more advanced lunar base, because temperatures are lower there. On the right are the green tubes of a 'hydroponic farm', where suitable vegetables provide not only food but oxygen, and there is an ore-processing plant. The scene is illuminated entirely by earthlight, the Earth appearing four times as big as the Moon appears in our sky, and up to 70 times brighter.

LUNAR PROBE: SMART-1

The first successful unmanned lunar probes were the Russian Lunas, launched as long ago as 1959. Many others have followed, and the entire surface of the Moon has been mapped in detail. As we move further into the future, it is likely that new technology will lead to designs different from anything forecast today. However, the illustration at left shows an actual vehicle, the European Space Agency's SMART-1, due for launch in the summer of 2003 at the time of writing.

SMART-1 was designed to be the first European spacecraft to orbit the Moon (all the earlier orbiters were Russian, US or, in one case, Japanese). It was also designed to be the first ESA vehicle to use electric or 'ion' propulsion. This accelerates a propellant – in this case xenon – to high velocity using electric power from solar panels. While producing much less thrust than chemical propellants, motors of this type can operate over long periods, building up to great velocity – essential for missions to deep space. It may well be that ion motors will become all-important in the future, so that much depended upon the performance of SMART-1.

The infrared spectrometer can map the surface distribution of minerals in more detail than ever before. One of the most important tasks for the probe is to peer into the craters near the poles; the floors of these craters are always in darkness and remain bitterly cold. The results from earlier US probes (Clementine and Prospector) gave rise to claims that ice could exist there. It is hoped that rays of reflected sunlight from the nearby crater rims will light up these floors sufficiently for ice to be detected – if it exists, which is very far from certain.

Extensive deposits of ice would be very helpful to the lunar colonists. Any lunar ice would, however, be mixed with rock-dust, and so would not be easy to extract.

The SMART-1 lunar probe. The European Space Agency (ESA) is testing new technologies which, if successful, will be used on much larger projects. One of the most promising of these is SMART-1 (Small Missions for Advanced Research in Technology), which uses a form of solar-electric propulsion known as a Stationary Plasma Hall-Effect Thruster, or PPS-1350, developed in France. Here we see the craft some 6,500km (4,000 miles) from the Moon, where it will go into orbit for some six months. Its next use by ESA could be for a Mercury mission called Bepi-Columbo, where it could reduce journey time by four years.

Overleaf:
Lunar exploration. Here we see astronauts who have discovered what appears to be an extinct thermal vent near the lunar pole – a sign of past volcanism, as also evidenced in the foreground rocks.

MARS IN THE 1950s

Mars has always been of special interest to us, because it is more like the Earth than any other world in the Solar System, a fact that gave rise to expectations it might support life.

Mars is much smaller and less massive than the Earth; its gravitational pull is only one-third Earth's. Its atmosphere is very thin. Yet only a few decades ago it was widely believed that advanced lifeforms might exist there.

Mars's two tiny satellites, Phobos and Deimos (named after the attendants of the mythological War God), are quite different from our massive Moon. The main belt of asteroids lies beyond Mars, and it seems that these two satellites were captured from there by Mars long ago; they are certainly asteroidal in character. No doubt they will eventually be pressed into service as natural space-stations.

In the 1950s it was believed that Mars's polar caps were made up partly of water ice but mainly of 'dry ice' – solid carbon dioxide. The wide tracts which give Mars its colour were regarded as deserts – not of sand, but of reddish minerals. The darker areas, described as greenish in hue, were thought to be old sea beds possibly filled with vegetation. Dust storms are common on Mars, and one famous astronomer, E.J.Öpik, maintained that these furnished proof of vegetation: Mars should appear featureless because of everything being buried under layers of dust, but of course vegetation can grow to penetrate the dust. It was also believed that, when a polar cap melted in the Martian spring, water vapour would be released into the atmosphere and the vegetation would thus be revived – so a 'wave of darkening' would spread from the pole down towards the equator.

Who has not heard of the 'canals' of Mars? In 1877 the Italian astronomer G.V. Schiaparelli made drawings of Mars showing thin, regular lines which he

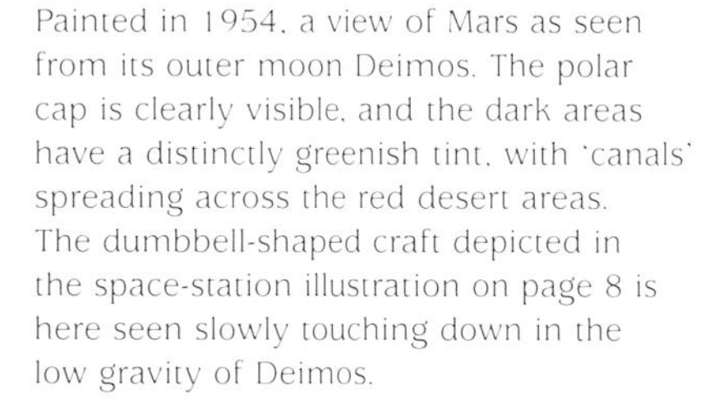

Painted in 1954, a view of Mars as seen from its outer moon Deimos. The polar cap is clearly visible, and the dark areas have a distinctly greenish tint, with 'canals' spreading across the red desert areas. The dumbbell-shaped craft depicted in the space-station illustration on page 8 is here seen slowly touching down in the low gravity of Deimos.

called '*canali*', or channels. Inevitably the Italian word was translated as 'canals', and the US astronomer Percival Lowell believed them to be artificial waterways, constructed by intelligent Martians as a planet-wide irrigation system to draw water from the polar icecaps. Lowell's maps, drawn from observations made using the large telescope at his observatory in Arizona, showed the canal network clearly – and, if those drawings were only accurate, there would be no question but that Mars is inhabited.

By the 1950s the idea of an advanced Martian civilization had been abandoned, but it was still thought the canals were narrow channels flanked on either side by strips of low vegetation.

It is now known that the canals do not exist in any form; they were simply tricks of the eye. Mars cannot support advanced lifeforms; the main problem now is to determine whether the planet supports any life at all.

The canals of Mars. Shown here are strips of vegetation nurtured by water from the melting polar caps. In the 1950s, the landscape of Mars was generally believed to be flat, with no mountains and at best rolling hills, because no long shadows could be seen telescopically at the terminator – unlike the case with the Moon, whose mountains cast such shadows. Hardy did, however, depict craggy buttes left by erosion, and this prediction has proved quite accurate.

When the 1972 version of this book was published, Mariner 9 was in orbit around Mars but Viking was still in the future. The two Viking probes were launched in 1975 to reach Mars in 1976. They achieved their objectives admirably, each Lander being released from its Orbiter (as shown here), which remained in orbit as a communications relay, also taking its own pictures.

MARS IN THE 1970s

The problem for telescopes of the 1950s was that Mars never comes much within 55 million kilometres (35 million miles) of us, and they could never show it more clearly than, say, a view of the Moon with good binoculars. Not until 1965 did the first successful Mars probe, Mariner 4, fly past the planet. One feature it recorded is visible with Earth-based telescopes as a tiny speck; named Nix Olympica, 'the Snows of Olympus', it had been assumed to be a snowy peak. Only when Mars could be seen from much closer range was it found that the feature is in fact a giant volcano, three times the height of Everest. It is now known as Olympus Mons – Mount Olympus – and is believed to be the highest and most massive volcano anywhere in the Solar System. Whether it is extinct, dormant or even mildly active is a matter for debate.

All our ideas about Mars had to be drastically revised in 1971, the year we had our first views of the huge volcanoes and the systems of canyons. Then came the two Vikings, launched in August 1975 and reaching Mars in mid-1976. One of their main tasks was to search for life, and this involved making controlled landings; the Lander was separated from the Orbiter and brought down to the surface, slowed partly by rocket braking and partly by parachute; tenuous though it is, with a ground pressure below 10 millibars everywhere, the Martian atmosphere is substantial enough to make parachutes useful.

Both landings were successful. Viking 1 came down on 10 June in the 'Golden Plain' of Chryse, 20° north of the equator, and Viking 2 landed on 7 August in the more northerly plain of Utopia. They sent back excellent images, relayed by the Orbiter. Rocks were everywhere. The sky was yellowish-pink, rather than the dark blue that had been expected. Wind-speeds were gentle. Temperatures were of course very low – far below freezing point. One NASA investigator, Garry Hunt, produced a wry weather forecast for Mars: 'Fine and sunny; very cold; winds light and variable; further outlook similar.' Not surprisingly, he proved to be completely accurate! Dust storms do occur, and can be global, but in general the atmosphere is very clear.

Each Lander was equipped with a 'grab' – basically, a scoop with a movable lid which could collect surface material and draw it back into the main spacecraft, where it could be analysed in what was to all intents and purposes a tiny but highly efficient laboratory. There were three experiments, all designed to detect biological activity. The results were decidedly puzzling, but in the end they were generally regarded as negative. The consensus was that there is no firm evidence for Martian life of any kind.

Throughout the 20th century, all the useful results from Mars missions have come from US spacecraft. The Soviet Union launched its first Mars probe as early as 1961, and others followed, but even today the Russians have had no success; all their Mars spacecraft have failed for one reason or another. This is all the more surprising in view of their excellent results from Venus, which logically would be expected to pose far more difficult problems.

Obviously, manned flight to Mars must be many orders of magnitude more hazardous than a trip to the Moon. The astronauts must endure months of

Right:
Again based on 1970s NASA plans, this shows the Martian Excursion Module (two in fact), together with a temporary base constructed from two pressurized half-cylinders. Another cylinder is making a soft landing using retro-rockets and a parachute. A large high-gain radio antenna is aimed at Earth, near the Sun in the sky, and at right is a Martian Rover. For the 1978 edition, *The New Challenge of the Stars*, elements of this painting were used, but, because of information from the 1976 Viking landers, with the sky now salmon-coloured.

Below:
The 1972 edition contained this painting of a Martian polar expedition, with a caption stating that this might take place before the end of the 20th century! The whitish deposits were considered to be mainly of carbon dioxide, rather than water, and the sky was a very dark, purplish-blue. The distant astronaut is using a sort of rocket-propelled pogo-stick, with a range of up to 30km (20 miles). For the 1978 edition, only the sky was altered – to orange, as in this 'split' view.

weightlessness and there is, perhaps above all, the danger from radiation; to provide adequate protection from radiation on a spacecraft is very difficult indeed. Neither are we yet sure whether the thin Martian atmosphere will be of any real use as a radiation screen.

Yet the Viking results were encouraging enough for NASA to press ahead with a design for the first Martian base. Though primitive, such a base would be essential, because the astronauts would have to spend some months on Mars before the two planets were suitably placed for the return journey to Earth. Walking about in the open with no protection apart from warm clothing and an oxygen-cylinder, as envisaged only a few decades earlier, was known to be out of the question because of the unexpectedly low atmospheric pressure; blood would boil inside the body, causing a quick but unpleasant death. Full pressure-suits would have to be worn all the time, and, naturally, any base must be very effectively airtight and pressurized.

In the 1970s it was still thought that the Martian polar caps were very thin, possibly no more than a surface layer of solid carbon dioxide. The scene shown here gives the impression of what a polar zone was thought to be like; only much later was it found that the caps are thick, and made up of water ice. We are also sure that even away from the poles there is a great deal of underground ice, so future colonists will never be short of water. In this respect Mars is much more cooperative than the Moon.

In 1972 it was thought that the first manned expeditions might set off before the end of the 20th century. This did not happen, but it will indeed be strange if attempts are not made during the next few tens of years, and these will lead on eventually to permanent bases. It is very likely that the 'first man on Mars' has already been born!

THE CANYONS OF MARS

Earth geology is dominated by studies of what is known as plate tectonics. The Earth's crust is divided up into definite plates that slide around on the partially melted mantle below. It was Alfred Wegener who in 1915 pointed out rigorously that the continents can be 'fitted together' as if they had been combined before the plates drifted apart. Others had spotted this 'jigsaw fit' before, but Wegener went further, tracing geological similarities across landmasses that are now widely separated. (For many years his theory of continental drift was ignored. Only since around 1960 has the science of plate tectonics been universally accepted.)

The theory of plate tectonics explains why volcanoes do not remain active indefinitely. A volcano arises over a 'hot spot' in the mantle, and will become extinct when the crustal drift carries it away from the hot spot. The main Hawaiian volcanoes demonstrate this well. Mauna Kea has drifted away from the hot spot, and so has ceased to erupt. Its place has been taken by the very active Mauna Loa.

On Mars, plate tectonics do not operate: when a volcano is formed over a hot spot it remains there. It can thus grow to immense size – as Olympus Mons, 40km (25 miles) high, has done.

There are two main volcanic regions on Mars. One is Elysium, in the northern hemisphere, but more spectacular is the Tharsis area. Olympus Mons is the greatest of the Tharsis volcanoes, though several others are comparable (Arsia Mons, Pavonis Mons and Ascraeus Mons, all much loftier than Everest).

Adjoining Tharsis are two magnificent systems of canyons that in the future are certain to become popular tourist attractions. Noctis Labyrinthus was seen by Earth-based telescopes, and was nicknamed the Chandelier, because of the arrangement of what we now know to be canyons (they were once thought to be watercourses, and the area was known as Noctis Lacus). The main canyon system is the one associated with the Valles Marineris and its tributaries. The valley itself is a huge gash in the surface, over 4,300km (2,700 miles) long, in places 600km (380 miles) wide and more than 6.5km (4 miles) deep. Compared with this, the Earth's Grand Canyon of the Colorado seems very puny.

Unlike many features that look like dry riverbeds, it does not seem that the Valles Marineris was cut by running water; it is too extensive, and too complex. It is unlike anything else on the surface, and shows that Mars, so calm and inactive today, must once have been a world in violent turmoil.

Spacecraft are now being sent to Mars at regular intervals. Perhaps the most successful have been the American rovers Spirit and Opportunity, which have moved around and have sent back magnificent images and data. We do not yet know whether there is any trace of life there, but we will undoubtedly find out in the near future – and it will indeed be surprising if men do not reach the planet during the first half of the present century.

Facing page:
Canyon on Mars. Tributaries to the main Valles Marineris canyon system, such as the Noctis Labyrinthus, may make spectacular viewing, especially when filled with dawn fog, as here, formed when water which during the night condensed and froze on east-facing slopes is vaporized by the rising Sun. Ice crystals in the high cirrus clouds form 'sun-dogs' around the early-morning Sun.

Below:
A composite of Viking orbiter images of Mars, showing the full extent of the gigantic canyon system Valles Marineris and its tributaries, as it would be seen from a distance of about 2,500km (1,500 miles). Noctis Labyrinthus is on the right. *(Photo courtesy NASA.)*

Overleaf:
Manned exploration of Mars. Here we see a Mars Rover equipped with grappling arms to allow the astronauts to gather larger samples than they could unaided. This vehicle can travel over a variety of terrains; seen here are a rock field, dunes and hills.

Note that no national emblems are shown in any of these illustrations. In the future such missions will hopefully be international; such countries as Japan, China and India are currently entering the arena of spaceflight.

MARS 3

TO THE MARTIAN POLE

The first bases on Mars are likely to be established in fairly low latitudes, where the temperatures are less extreme than in other parts of the planet. Chryse Planitia, site of the first Viking landing, is one favoured location, but much will depend on what we find out before the main expeditions are ready; in particular, are there ponds or even lakes of liquid water not far below the surface? The various probes dispatched during the first two decades of the 21st century should be able to answer this question.

There are various ways of exploring the less accessible areas, some of which must be of exceptional interest. For instance, some small craters have gullies in their walls which give every impression of having been water-cut, and which are not eroded or dust-filled, so can hardly be very old. Wheeled vehicles can of course be used, and automatic rovers were sent to Mars even before the end of the 20th century. A more startling proposal is to set up a railway system; when colonization of the planet is well under way there should be no problem in laying tracks across the surface, and the engines could be solar-powered. Eventually, when there are bases in various parts of Mars, there will no doubt be an efficient railway network.

Aircraft will be different from those of Earth, because of the thinness of the atmosphere – the pressure is no greater than that of the Earth's air at a height of about 37,000m (120,000ft) above sea level. Yet a suitably designed aircraft would be able to function. A prototype, called the *Ares*, was designed by NASA in 2003, although it has yet to be built; testing it on Earth will be difficult, because of the much greater gravity here. Even so, aircraft will almost certainly have a role to play in Mars's future.

Overland exploration of Mars will mean designing suitable vehicles, but this

This painting, *Departure for the Pole!*, was commissioned in 2000 by the 2111 Foundation for Exploration (now the Earth and Space Foundation). It depicts an expedition leaving for a transpolar assault on Mars's north geographical pole. The expedition is setting off just before sunrise to get a day's traverse. Members of the temporary base camp at the edge of the polar cap wave goodbye to the explorers. In view are two of the skidoos and their cargoes (there may be more to the right of the picture). Our imaginations are left to contemplate the scene in front of the expedition team – the vast expanse of the polar plateau. Note also the 'layered terrain' behind the base.

Charles Cockell wrote: 'This painting is a significant work as it was the first painting to be rendered by an artist of an overland expeditionary attempt on the Martian poles.'

should be no problem. One expedition will certainly go to the north pole, one of the coldest and most inhospitable regions on the entire planet – though the temperatures are never so low as they can be at the south pole, because in one respect Mars differs markedly from the Earth. Its orbit is much more eccentric, and the distance from the Sun ranges from 206,500,000km (128,400,000 miles) at closest approach (perihelion) out to 249 million kilometres (154,860,000 miles) at furthest recession (aphelion). This means that at perihelion Mars receives 44 per cent more solar radiation than it does at aphelion.

As with Earth, on Mars summer in the southern hemisphere occurs near aphelion. This means that, in theory, for both planets the climates in the southern hemisphere should be more extreme than those in the north: the summers shorter but hotter, the winters longer and colder. With Earth this is not pronounced, however, because our orbit is so nearly circular, and moreover there is much more ocean in the south to act as a stabilizing influence. Not so on Mars, where the differences between the two hemispheres are very marked. The south polar cap can become much larger than its counterpart in the north, but at minimum extent, during the southern-hemisphere summer, becomes very small.

The axial tilt of Mars is almost the same as the Earth's, so the seasons are of the same type – although much longer, for the simple reason that the Martian year is much longer.

There is another important effect. Because of the presence of the comparatively massive Moon, the axial inclination of the Earth hardly varies over the millennia. Mars does not have such a check on the shifting of its axial tilt; Phobos and Deimos are much too small to have any detectable effect. The result is that the tilt ranges between 15° and 35.5° over a cycle of 51,000 years. If the south pole is tilted sunward in summer when the inclination is greatest, it may well receive enough heat to cause much of the ice to sublime (change directly from solid ice to water vapour), temporarily thickening the atmosphere and even producing rainfall. It has also been suggested that Mars may occasionally go through spells of violent volcanic activity, when tremendous quantities of gases and vapours, including water vapour, are sent out from beneath the crust.

Long-term changes of this kind would be particularly easy to see in the polar regions, where layered terrain has already been detected. Moreover, it is quite possible that we are now seeing Mars at its most unwelcoming, and that in the future warmer, wetter conditions will return – whether or not we contribute to the process. Much is bound to be learned from the first overland journeys to the Martian pole. They should be undertaken not long after the first permanent base is established on the surface.

Mars today is a world where we could not live except under very artificial conditions. But could we terraform it – that is, change it into a sort of second Earth?

Two main things have to be done if we're to terraform Mars: thicken and modify the atmosphere, and raise the temperature. One suggested method is to introduce 'greenhouse gases' to shut in the Sun's heat. Another is to coat the polar icecaps with dark material so that more infrared radiation is absorbed, releasing locked-up water and increasing the density of the atmosphere. Bacteria 'seeded' in the atmosphere and, next, planted vegetation would produce oxygen. The process would be slow; at best, terraforming Mars will take centuries, should it even prove practicable at all. Carl Sagan once suggested crashing an icy asteroid onto the surface, but this would seem to be somewhat drastic – as well as highly dangerous to any colonists who might be there at the time!

Overleaf:
Mars as seen from its small, carbonaceous satellite Phobos. But this scene is of a far-future Mars, when the planet has been terraformed: it has oceans, and the Valles Marineris has become the first Martian canal.

THE OLD VENUS

Venus, the lovely Evening and Morning Star, is the most brilliant object in our entire sky apart from the Sun and the Moon. At its best it looks like a small lamp, and can cast perceptible shadows. In size and mass it is very similar to the Earth, and the two worlds have often been regarded as twins – but, as we have found, they are non-identical twins. Though Venus was named in honour of the Goddess of Love, it is anything but a friendly world.

Venus is on average 108 million kilometres (67 million miles) from the Sun, and has an orbital period of nearly 225 Earth days. As seen from Earth, it shows phases similar to those of the Moon, but telescopes show almost nothing on its disc because the actual surface is permanently hidden by a dense, cloudy atmosphere. The clouds never clear away; there is no such thing as a sunny day on Venus.

Before the Space Age, we knew nothing about the surface conditions, and even the axial rotation period was unknown. Venus was indeed the planet of mystery. Spectroscopic work showed that the upper atmosphere at least was rich in carbon dioxide; since this gas acts in the manner of a greenhouse, it was safe to assume Venus must be a hot world. Svante Arrhenius, a Swedish chemist whose work was good enough to earn him a Nobel Prize, believed Venus to be similar in condition to the Earth during the Carboniferous Period, around 350 million years ago, with luxuriant vegetation, extensive swamps, and no doubt lifeforms such as amphibians and insects. Other astronomers considered that there could be broad oceans, with relatively little dry land. In this case the atmospheric carbon dioxide would have fouled the water to produce seas of soda water. It is probable that life on Earth began in our seas, and the same would presumably be true of Venus, so life could evolve in the same way as it did here. Later, Fred Hoyle suggested there might be seas of oil.

However, spectroscopic analysis showed no trace of either oxygen or water vapour; so, according to a different theory, Venus was a bone-dry, fiercely hot desert. In this theory, what we were seeing were clouds of dust, dense enough to mask the surface completely. High winds would erode the rocks into strange shapes, like alien sculptures. To attempt a landing there would be very hazardous.

Why should Venus and the Earth be so different? Presumably the reason is that Venus is closer to the Sun. It seems fairly definite that the planets were formed, around 4.6 billion years ago, from the solar nebula, a cloud of material surrounding the youthful Sun. Initially Venus and the Earth might well have been similar, with oceans of water; extremely primitive, bacteria-type lifeforms could well have appeared on Venus. But in those far-off times the Sun was not so luminous as it is now. As it has aged, it has become hotter. The Earth was far enough out that it was not badly affected, but for Venus the situation was very different. The oceans boiled away, the carbonates were driven out of the rocks, and in a short time (by cosmological standards) Venus changed from a potentially life-bearing planet into the furnace-like world of today. Any life there would have been ruthlessly snuffed out.

But did life appear when Venus was cooler than it is now and had surface water? We cannot tell, but it seems certain life would have had little time to develop before the conditions became intolerable. It is hardly likely that any specimens brought back by future spacecraft will contain recognizable fossils.

In a way Venus has been a disappointment; conditions there are very like the conventional picture of Hell. It is sobering to reflect that, if the Earth had been only about 30 million kilometres (20 million miles) closer to the Sun, it would have suffered the same fate.

Facing page:
The surface of Venus – two possibilities imagined in the 1950s. Was Venus a watery world with primeval vegetation or an eroded dust-desert? We now know the answer: neither!

VENUS IN THE 1970s

By 1972 at least some of the mysteries of Venus had been solved. The first successful mission was the USA's Mariner 2; the probe flew past Venus in December 1962 and sent back reliable data. The idea of oceans was abandoned at once; the surface temperature was far too high. There was no detectable magnetic field.

The axial rotation period, previously thought to be 'a few weeks', was found to be very long. It is now known to be 243 days – longer than Venus's orbital period, leading to a most peculiar calendar. And the rotation is retrograde, or backward, compared to the Earth or Mars: to an observer on the Venusian surface, the Sun would rise in the west and, 118 Earth days later, set in the east. (In fact, the Sun could never be seen from the surface because of the permanent cloud cover; the only indication of its position would be a baleful glow in the sky.)

Why does Venus rotate in an east-to-west direction? In 1972 we did not know – and this is still true today. According to theory, Venus was struck by a massive body at an early stage, and literally tipped over. This does not sound very plausible, but it is difficult to think of any better explanation, and conditions in the early Solar System were much more chaotic than they are today.

Another peculiarity of Venus is that, although the planet's rotation period is so long, the upper clouds have a rotational period of only four days, again in an east–west motion.

The atmospheric structure of Venus is remarkable, and unlike anything else in the Solar System. Surface winds are gentle, and the temperature is almost 500°C (930°F). What happened is that the atmosphere, consisting mainly of carbon dioxide, created a 'runaway greenhouse' effect. This was known in 1972, when the first edition of this book was published. It was thought there would be lightning flashes, and that the surface might actually be red-hot. Moreover, as Carl Sagan and others pointed out, with such a dense atmosphere a phenomenon known as 'supercritical refraction' would come into play; an observer would seem to be standing at the bottom of a huge bowl, with the horizon curving upward on all sides. Just as a pencil standing in a bowl of water appears to be bent, so on Venus one should in theory be able to see the back of one's neck – if, of course, the atmosphere were clear . . .

By 1972 we knew the surface of Venus was even hotter than we'd supposed, due to a runaway greenhouse effect. The surface might glow, and there would be thermal activity and molten lava. By now, Venus had earned itself the title of Hell Planet.

In fact this does not happen, as was realized by the time *The New Challenge of the Stars* was published in 1978, pictures having been obtained direct from the surface. The USSR took the lead in Venus probes, and in 1975 their Venera 9 succeeded in sending back one good image (after landing, the spacecraft continued to transmit for 55 minutes before being put out of action by the intensely hostile conditions). The landscape was described as a 'heap of stones', sharp-edged and showing little sign of erosion. The light level was low, and was compared with that at noon in Moscow on a cloudy winter day. This was less gloomy than had been expected; Venera had been equipped with floodlights, but there was no need to use them.

Detailed maps of the surface have since been obtained using radar; of special note are the results from the Magellan Orbiter, which reached Venus in 1990 and remained in orbit until 1994. There are high mountains, deep valleys, volcanoes and impact craters; lava flows are everywhere, and the landscape is essentially volcanic. It is very probable that volcanic activity is currently widespread; significantly, the quantity of sulphur dioxide

LEFT:
Venus, 1978 version. Sunlight briefly penetrates the cloud layer to illuminate a balloon probe floating in the thick atmosphere above volcanoes – some of which may some still be active. The probe's instruments might eventually succumb to sulphuric-acid rain.

ABOVE:
The clouds of Venus, a painting based on images from Mariner 10 (1974) and the Pioneer Venus orbiter (1978). Through an Earthbound telescope, Venus is a white disc almost devoid of features, though occasional dark markings appear. Astronomers must use other wavelengths, such as ultraviolet, to see the actual structure and flow of the clouds.

(first detected in 1974 by the Mariner 10 spacecraft, en route to Mercury) in the atmosphere varies. We showed a hydrogen-filled balloon in the painting; because of the lack of oxygen there is no chance of an explosion, and a balloon of this kind could float around in the atmosphere for months. In 1985 two balloons were indeed dropped into the atmosphere by the Russian Vega probes, on their way to Halley's Comet, and they sent back useful data.

There are two main highland areas on Venus. Ishtar Terra is about the size of Australia, Aphrodite Terra rather larger. The plains cover about 68 per cent of the surface, with lowlands accounting for less than 30 per cent. Adjoining Ishtar are Venus's highest mountains, which rise to almost 11km (7 miles) above the mean radius of the planet.

To make the overall picture even less attractive, it has been found that the clouds, which look so beautiful, are rich in sulphuric acid. It is safe to say that the chances of a manned flight to Venus in the foreseeable future are effectively nil!

We now know that Venus does share many geological features in common with the Earth. The highland areas may be regarded as 'continents' and the low-lying areas resemble our ocean floors. The volcanoes are of the shield variety, as are those on Mars; apparently one huge crustal plate covers the surface (rather than the plurality of tectonic plates we find on Earth, relative motions of which give rise to earthquakes). Of special interest is the highland area of Beta Regio and Phoebe Regio, where we have what is partly a large shield volcano – very probably active – and partly highland cut by a rift valley.

Impact craters abound, but they differ from those on Mercury or the Moon, because the dense atmosphere acts as a screen. There are virtually no impact craters below 30km (20 miles) across, but there are some very large structures, well over 250km (150 miles) in diameter.

Much research remains to be done on Venus, and we may in the process learn a great deal about our own planet.

OVERLEAF:
The surface of Venus, a view based on modern radar observations and the results from many probes, both US and Russian. The landscape is basically volcanic, with huge lava-flows and wide shield volcanoes, similar to those on Hawaii but even larger. At the surface the pressure is 90 atmospheres; but at a height of 50km (30 miles) it is only one atmosphere (i.e., the same as at Earth's sea level), and the temperature falls to 30–80°C (86–176°F). The fact that sulphur dioxide and hydrogen sulphide co-exist here, when normally they would react and destroy each other, has led some scientists, like Dirk Schulze-Makuch, to suggest there might be micro-organisms living at this level, especially since there does not seem to be enough volcanic activity to explain the phenomenon.

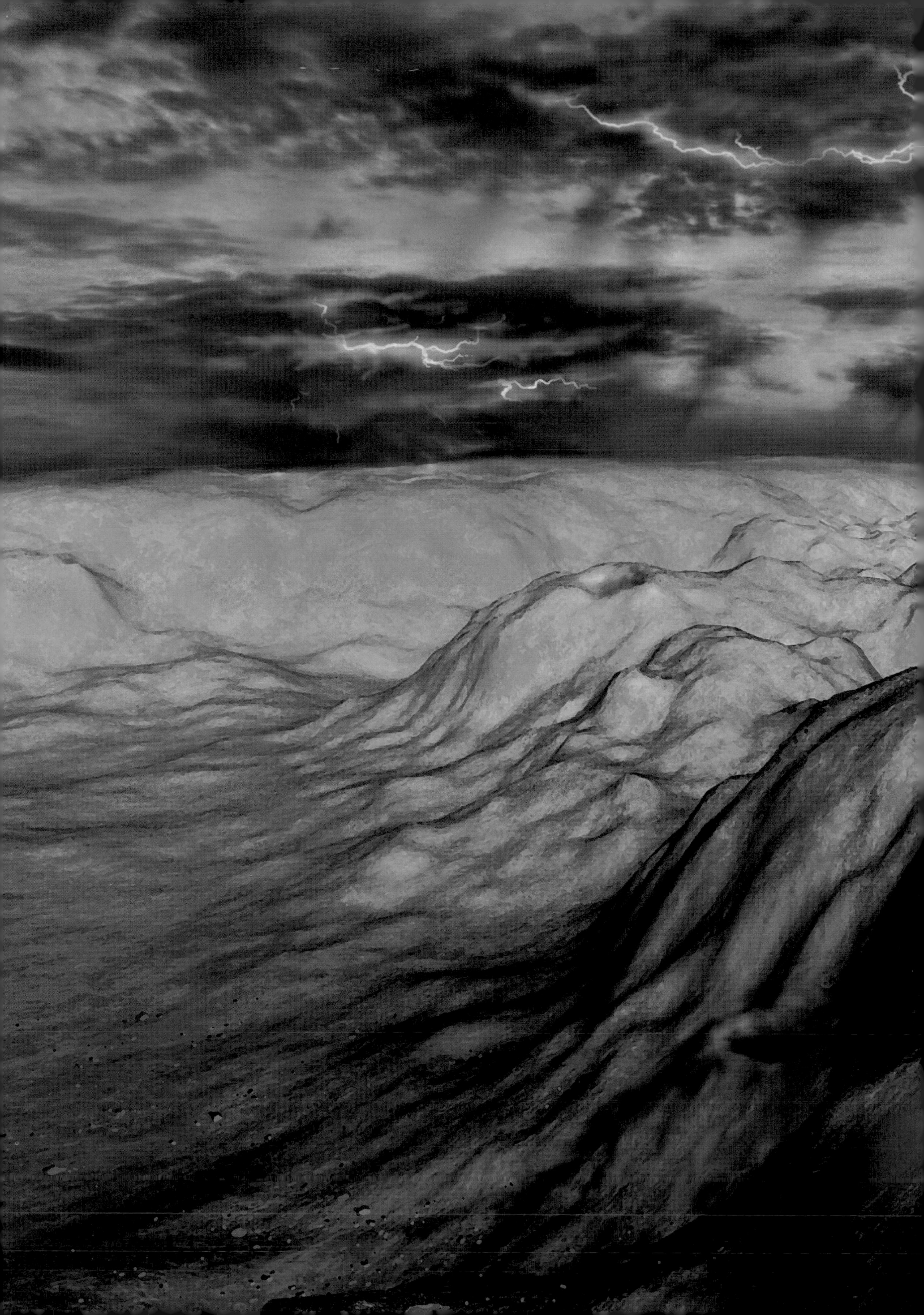

ABOVE:
This 1972 view of the surface of Mercury shows the Sun eclipsed by a plug of volcanic lava. The Earth–Moon system is visible as a double star, immersed in the Zodiacal Light (a band of debris left over from the formation of the Solar System).

FACING PAGE:
Mercury and the Sun, the latter with its corona and Zodiacal Light (neither corona nor Zodiacal Light would normally be seen except during an eclipse, and nor would the background stars, but artist's licence is at play here!), as they must have appeared to the Mariner 10 probe in 1974. It is hoped that NASA's MESSENGER – which stands for MErcury Surface, Space ENvironment, GEochemistry and Ranging Mission – probe (part of its Discovery Program of supposedly better, faster and cheaper missions) will at last map the whole planet.

MERCURY

Mercury, the closest planet to the Sun, is never very conspicuous to us, because it always stays near the Sun in the sky, and from Earth can never be seen against a really dark background. With the naked eye it is seen either low in the west after sunset or low in the east before dawn. Ordinary telescopes show virtually no features apart from the characteristic phase, because Mercury is a small planet – just over 4,800km (3,000 miles) in diameter – and never comes much within 80 million kilometres (50 million miles) of us. In size and mass it resembles the Moon much more closely than the Earth, and the little world's low escape velocity means there is only an excessively thin atmosphere.

From an artist's point of view, Mercury has not 'changed' as much as have most of the other planets – except in one important respect. Until the 1960s it was believed Mercury had synchronous rotation – that is to say, the rotation period equalled the orbital period (88 Earth days). One hemisphere of Mercury would be in permanent sunlight while the other would always be in deepest night. Mercury would be both the hottest and the coldest planet, with one side as hot as molten lead and the other only just above absolute zero. Between these two extremes would be a twilight zone where the Sun would rise and set, but always keep close to the horizon.

Then it was found that Mercury's dark side is not so cold as it would be if it never saw the Sun, and radar measurements confirmed that the planet's true rotational period is 58.6 Earth days – two-thirds of a Mercurian 'year'. To the disappointment of some science-fiction writers, there is no 'twilight zone' where the temperatures would be tolerable.

One spacecraft has flown past Mercury: Mariner 10, which first bypassed Venus to put it into the required path and then made three active passes of Mercury, on 29 March and 21 September, 1974, and 16 March, 1975. The minimum distance from Mercury during the last encounter was only just over 300km (200 miles). Unfortunately the same areas of Mercury were mapped each time, and over half the planet's surface remains unexplored.

The photographs sent back look very like views of the Moon, but closer inspection shows significant differences, including features not found on the lunar surface. There are thrust faults, ridges (such as Discovery Scarp) and lightly cratered areas; there are also 'lobate scarps', high cliffs that seem to be in the nature of thrust faults cutting through other features and displacing older landforms.

The most imposing feature is the Caloris Basin, so called because, every two years, when Mercury is at perihelion, the Sun is directly overhead there. Caloris Basin is an impact crater over 1,250km (800 miles) in diameter, and is bounded by a ring of smooth mountain blocks. Its floor is covered with a crazy-paving pattern of ridges and clefts. On the exact opposite side of Mercury there are peculiar, corrugated hills, probably caused by the focusing of seismic waves generated by the massive impact that created Caloris Basin.

Mercury is probably less hostile than Venus, but any form of life there seems to be out of the question. A robot transmitting station may well be set up there, but the first manned flight is likely to be long delayed.

ASTEROIDS AND COMETS

Asteroids and comets are the junior members of the Solar System. Though they all move round the Sun like the Earth and the other planets, some of them have orbits which are highly eccentric and inclined. The main belt of asteroids lies between the orbits of Mars and Jupiter, and was once regarded as potentially dangerous to spacecraft passing through en route for the outer planets. However, so far there have been no problems on this score.

Several main-belt asteroids have been imaged from close range, and not all are alike. All are cratered, but some have smooth areas; some are reflective, while others, such as Mathilde, are as black as charcoal. Small asteroids are often irregular in outline, and are certainly fragments of larger bodies that have been broken up by collision. Asteroid No. 216, Kleopatra, is shaped very like a cartoon dog's bone!

Few main-belt asteroids are more than 250km (150 miles) across. Much the largest is Ceres, 968km (602 miles) in diameter. The only asteroid ever clearly visible with the naked eye is the 530km (330 mile) Vesta.

Asteroid No. 433, Eros, has an orbit that brings it closer in than the main belt; it can occasionally approach the Earth to within about 25 million kilometres (15 million miles), as it did in 1931 and again in 1975. It is shaped rather like a bread roll, with a longest diameter of a little over 30km (20 miles), and rotates in a period of just over five hours. Its gravitational pull is very feeble, so that landing there will really be more a docking operation. This 1972 painting shows astronauts who have just docked, and have set up an inflatable, pressurized dome in preparation for a geological survey. Their spacecraft is almost over the horizon; Mars shines through the glow of the Zodiacal Light, and the surface of Eros is shown pitted with tiny craters, caused by impacting debris.

In fact, an unmanned spacecraft made a controlled landing on Eros in February 2001; this was NEAR – Near Asteroid Rendezvous Vehicle – later named in honour of the US astronomer Eugene Shoemaker. The landing site was inside a 10km (6-mile) saddle-shaped structure, Himeros; the daytime temperature was found to be 100°C (212°F), plunging to below –150°C (–240°F) at night.

Eros is a very ancient body, formed in the early history of the Solar System. It is a monolith, composed mainly of iron, magnesium and nickel. Many craters were found, together with scattered boulders and 'ponds'. It is thought that most of the larger rocks were ejected from a single meteoritic impact perhaps a billion years ago.

Eros is only one of a number of asteroids known to swing closer in than the main belt. These are known as Apollo asteroids, and some may approach the Earth, giving cause for concern. There are also asteroids whose paths take them from the main belt

This 1972 painting shows astronauts who have docked with Eros (rather than landed on it, because of its low gravity), a 27km x 15km (17 mile x 10 mile) asteroid shaped rather like a bread roll, which rotates in five hours, and which can approach Earth within 24 million kilometres (15 million miles). They have set up an inflatable, pressurized dome in preparation for a geological survey. Their ship is almost over the low horizon. In the plane of the Zodiacal Light, Mars can be seen. Eros's surface is shown pitted with tiny craters caused by the impacts of smaller debris within the asteroid belt.

An imaginary comet probe. The crescent Earth and Moon reveal the direction of the Sun, away from which a comet's tail always points. The original painting was purchased by Carl Sagan.

to within the orbit of Mercury; at perihelion they must be red-hot, but they become bitterly cold when near aphelion, so their climates must be decidedly uncomfortable. Two examples are 1566 Icarus and 3200 Phaethon. Icarus is less than 1.5km (a mile) in diameter.

Comets are quite unlike planets. They are of very low mass, though some of them can become very spectacular. They move around the Sun, but most have very eccentric and inclined orbits. Many comets are known to have short orbital periods (only 3.3 years in the case of Encke's Comet), and we can always keep track of these, even though most are very faint. The only conspicuous comet with a period less than two centuries is Halley's, whose period is 76 years; the last return was that of 1986.

The truly brilliant comets have much longer periods, so their visits cannot be predicted. Comet Hale-Bopp, visible with the naked eye between July 1996 and October 1997, will not be back for about 2,400 years.

Short-period comets come from the Kuiper Belt (see page 73), a region beyond the orbits of Neptune and Pluto, while it seems that longer-period comets come from the Oort Cloud, a swarm of icy objects over a light-year away.

Many comets, including Halley's, move in retrograde orbits. Their low masses mean their paths can be easily perturbed by the planets, and in some cases this perturbation may be enough to eject them from the Solar System; others may be diverted to plunge into the Sun.

The only fairly substantial part of a comet is the nucleus, made up of rocky particles held in ices; even Hale-Bopp, a giant by cometary standards, has a nucleus less than 50km (30 miles) across. When a comet moves in towards the Sun and is warmed, the ices begin to evaporate and the comet develops a tail or tails. Tails are of two types: those made up of gas and those made up of 'dust'. The effects of sunlight and the 'solar wind' mean cometary tails always point away from the Sun; when a comet is moving outwards, after having passed perihelion, it travels tail-first.

Several comets have now been surveyed by spacecraft, but when we wanted to depict such a scene in 1972 we had to invent one. Since the probe was supposed to have passed through the tenuous coma (head) of the comet, there should perhaps be more evidence of impacts by solid particles. The first actual encounters did not take place until 1986, when no less than five probes were sent to Halley's Comet. Of these, two were Russian, two were Japanese and one was European.

The European probe, Giotto, went into the comet and imaged the nucleus from a range of only 1,675km (1,040 miles); transmissions stopped following collision with a particle no larger than a grain of rice; the impact caused Giotto to gyrate. By this time, though, the camera had already sent back images of a potato-shaped body, measuring 13km x 8km (8 miles x 5 miles), with crater-like hollows and 'mountains'. Jets of gas were coming from the warmed, sunlit side. Material making up its tail is permanently lost to a comet; in the case of Halley's Comet, about 300 million tons of material is lost at each return to perihelion.

Overleaf:
The cratered and pitted surface of a small asteroid which has come within 100,000km (63,000 miles) of the Earth, which planet is seen eclipsing the Sun, its atmosphere forming a 'ring of fire' (the redness is caused by sunlight being refracted by the atmosphere, as at sunset). The Moon is on the right. While such a body *might* pass us by, there is every chance that Earth's gravity would draw it into a collision . . .

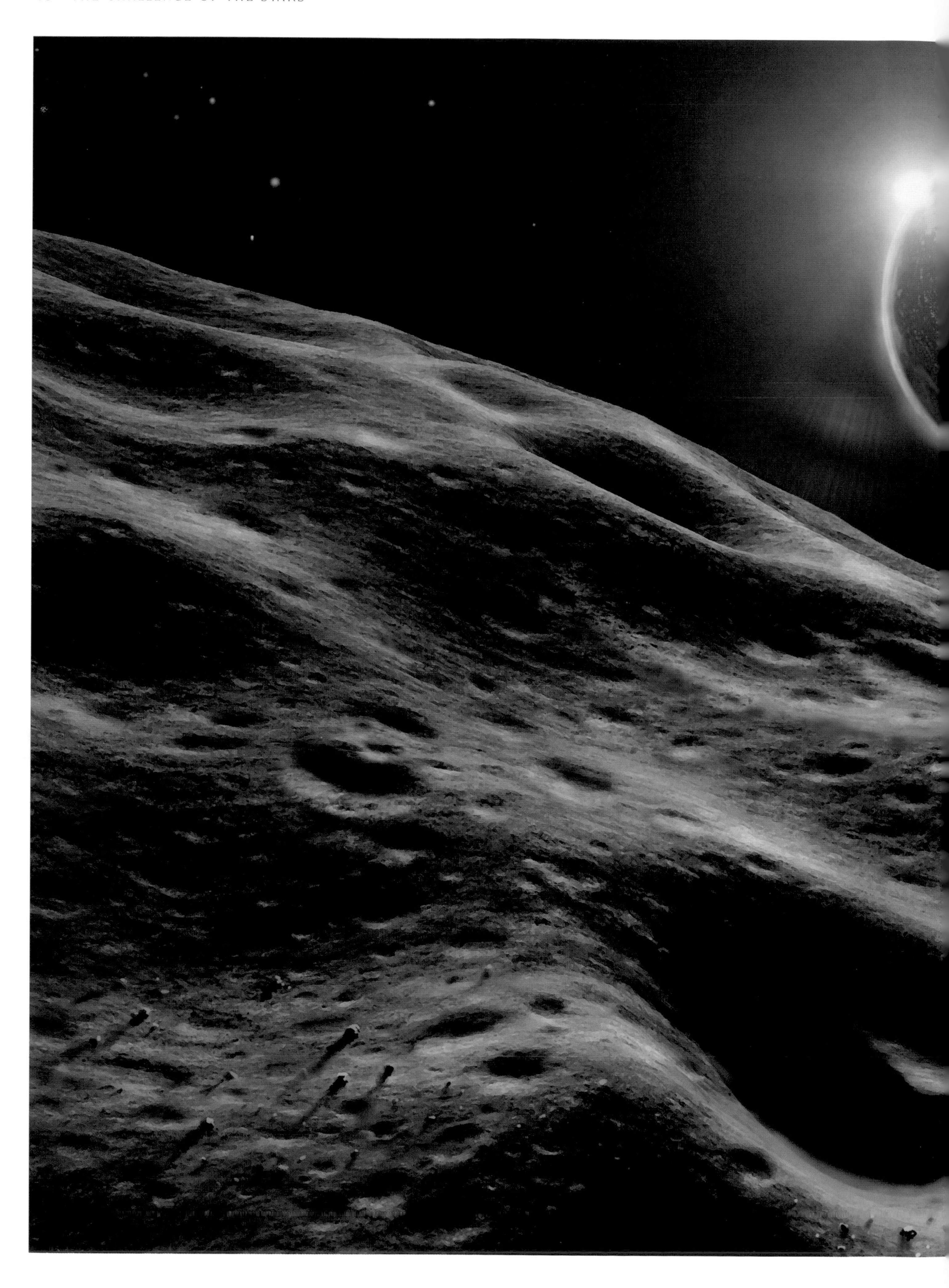

ASTEROID IMPACTS

Near-Earth asteroids (NEAs) are now known to be much more common than used to be believed was the case. During the past few years, several have been observed passing between the Earth and the Moon, though it is significant that they escape detection until they have passed their closest point of approach. All of these have been very small, but one asteroid whose orbit crosses that of the Earth – No. 1036, Ganymed – is as much as 40km (25 miles) in diameter. (Do not confuse this asteroid with Ganymede, the largest satellite of Jupiter. It is unfortunate that the names are so alike.)

The chances of the Earth being hit by an asteroid massive enough to do substantial damage are slight, but they are not nil. Indeed, an event of this sort may well have happened about 65 million years ago. The Earth was hit at that time by what is believed to have been an asteroid or cometary nucleus perhaps 30km (12 miles) in diameter, and the site has been identified: Chixulub, on the Yucatán platform off the coast of Mexico, where there is a 240km (150-mile) crater partly submerged beneath limestone. Layers laid down there contain iridium, which is normally very rare but is characteristic of meteorites. Huge waves must have swept over the land, and heated debris rained down after the collision, starting numerous fires. The Sun was blotted out, and the entire world climate changed – with fatal results for the dinosaurs, then the most advanced creatures on Earth. We cannot be one hundred per cent sure that the extinction of the dinosaurs was really due to this impact, but it does seem very probable.

Much larger, earlier impacts may well have sterilized the surface of the Earth, and in the case of comets have introduced large quantities of water. They may also have brought organic materials, and it has even been suggested that life on Earth was brought here by an impacting body, though the various 'panspermia' theories, of which this is one, have never met with a great deal of support.

The cratered surfaces of the Moon, Mercury, Mars, many of the planetary satellites and the larger asteroids confirm the great abundance of such small bodies thousands of millions of years ago. In many cases, notably that of the Moon, impact craters dominate the entire scene. On our planet, the atmosphere, the oceans and geological processes like volcanism and plate tectonics have obliterated the evidence to a great extent, but many examples of impact craters have been identified, and some of them are striking – no pun intended! The most famous example is the Barringer Meteor Crater in Arizona, formed about 50,000 years ago and described by the Swedish scientist Svante Arrhenius as 'the most interesting place on Earth'. It is over a kilometre (over 4,000ft) in diameter and about 230m (750ft) deep. Another well preserved impact crater, around 300,000 years old, is at Wolf Creek in Australia.

Although the impact of a body as large as the one shown in this painting is unlikely in the near future, it is by no means impossible, and even a much smaller impacting asteroid would do immense damage. It would be as well to have advance warning, which is why the Spaceguard project was set up by the late Dr Eugene Shoemaker (whose name was given to Comet Shoemaker-Levy 9, which struck Jupiter in 1994 with dramatic results – easily observed from Earth). Some organizations try to convince governments of the importance of a search for possible Earth-impactors, but as yet there are only six detection programmes in operation worldwide: five in the USA and one in Japan. The European Space Agency has a project called SIMONE to send five probes (using ion-thruster technology similar to that of SMART-1; see page 14) to different types of NEA to transmit back information about their composition and potential for damage. Britain has Spaceguard UK, based at Knighton in Wales. There are no comparable programmes in the southern hemisphere, though of course the threat could as easily come from this direction.

If there is sufficient notice, attempts could be made to divert the body by a nuclear explosion, and it might also be practicable to evacuate the threatened areas.

Facing page:
A large asteroid strikes an inhabited region of Earth – the worst-case scenario. Even if advance warning was given, it's evident that any measures taken to destroy or deflect the asteroid must have failed. Perhaps attempts were made to evacuate the impact area; since this is the night side of Earth we can see not only the lights of great cities but the fires caused by a panicked populace. Yet the truth is that there's nowhere to run; an impact as large as this one would create a 'nuclear winter' effect as vast quantities of dust, smoke and debris were hurled into the upper atmosphere, blotting out the Sun's rays.

SURFACE OF A COMET

A cometary nucleus consists mainly of water ice and rocky particles, but it is significant that the first nuclei to be examined by spacecraft, those of comets Halley and Borrelly, were found to be as black as coal due to a coating of hydrocarbons. Jets of gas erupt from below, and there is a constant outflow of particles that are slow-moving but which could well endanger any approaching lander. In our painting the lander, firing its retro-rockets, is shown eclipsing the Sun.

NASA planned to launch a probe, Rosetta, in 2003 to rendezvous with Comet Wirtanen, which has an orbital period of 5.5 years. Put into a path around the comet, Rosetta would have released a small lander to make the first controlled touchdown on the surface of a comet. Rosetta should have been sent up from Kourou, in French Guiana, on an Ariane 5 rocket, but problems with the Ariane meant the launch window was missed. Rosetta was launched in 2004, and, if all goes well, will rendezvous with a larger comet, Churyumov-Gerasimenko, in 2014.

The first space-craft to hit a comet was the American Deep Impact, which on 4 July 2005 crashed upon the periodical comet Tempel 2. The copper impactor had been on its way for 172 days. The impact velocity was 10 kilometres per second; there was a bright flare, and the comet brightened by two magnitudes. The impactor was of course vaporized, but there was no effect upon the motion of the comet.

A large crater was formed in the pear-shaped nucleus of the comet, which was about 5km across. The encounter was imaged by the orbital section of the space-craft, and observations were made from Earth. The main surprise was that the ejecta cloud was made up not of water, ice and dirt, but of very fine, powdery material.

After the encounter, Tempel 2 soon returned to normal, and was completely undamaged apart from having acquired a new crater.

The surface of this comet looks bright only because of the glare of sunlight (even the Moon is dark in reality) and because we are seeing an area which has been disturbed by evaporation and the outgassing of volatiles, revealing ices. In the foreground of this scene are dark rocks perched on pinnacles of ice, which they have shielded from sunlight.

JUPITER

Jupiter, the giant of the Solar System, is more massive than all the other planets combined. It moves around the Sun at a mean distance of 777 million kilometres (483 million miles) in a period of 11.9 years. In our skies it is a brilliant object; despite its great distance it is outshone only by the Sun, the Moon, Venus and (very occasionally) Mars. Its axial rotation period is less than ten hours, and this explains why its disc is so obviously flattened. The equatorial diameter is over 143,000km (89,000 miles), the polar diameter about 133,000km (83,000 miles).

The structure of Jupiter is quite unlike that of the Earth. According to modern models, there is a hot silicate core; the temperature here is at least 20,000°C (36,000°F), probably rather more. Above this is a thick shell of liquid metallic hydrogen, which is itself overlaid by a shell of molecular hydrogen. Finally comes the gaseous atmosphere, about 1,000km (600 miles) deep and made up mainly of hydrogen and helium. The upper clouds are bitterly cold – about –150°C (–240°F). Telescopes show that the yellowish disc is streaked with cloud belts. The Great Red Spot, which has been under observation since the 16th century, is now known to be a whirling storm – a phenomenon of Jovian 'weather'.

Four of Jupiter's satellites are of planetary size: Io, Europa, Ganymede and Callisto. Ganymede is actually larger than Mercury, though less massive. These satellites are known collectively as the Galileans, since they were observed by Galileo as early as 1610. All the remaining satellites are small; only Amalthea and Himalia are as much as 160km (100 miles) across.

The planet and its major satellites have been studied in detail by the various Jupiter spacecraft, notably the Pioneers, the Voyagers and Galileo. They are not alike. Ganymede and Callisto are icy and cratered, Europa is icy and smooth, and Io is violently volcanic.

This painting, from the 1972 edition of *The Challenge of the Stars*, is a good example of how the results from the space probes have revolutionized our ideas about the planets and their satellites. The original caption read: '. . . the mountains, which are covered with "ice" – not ordinary ice, but gases which have been frozen; the temperature is below -200° Fahrenheit . . .' But there are no mountains on Europa, as we found to our surprise when the Voyager probes sent back close-range images in 1979. Relative to its size, Europa is smoother

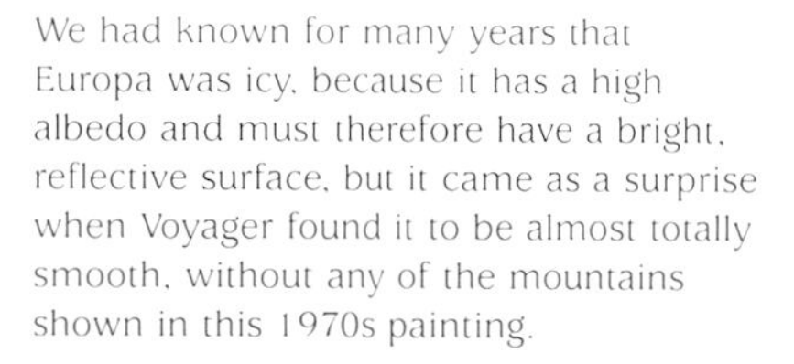

We had known for many years that Europa was icy, because it has a high albedo and must therefore have a bright, reflective surface, but it came as a surprise when Voyager found it to be almost totally smooth, without any of the mountains shown in this 1970s painting.

Believed in 1972 to be the closest satellite to Jupiter, orbiting only 113,000km (70,000 miles) from the cloud-belts of the giant planet, Amalthea – or Jupiter V (because it was the fifth moon to be discovered, in 1892) – seemed would make an ideal site for a manned observatory. The one shown here was to have been landed prefabricated by means of a propulsion module. But we now know that Jupiter's intense radiation belts make this a dangerous site for any such project! *From the private collection of Adrian Berry*

than a billiard ball. The surface is cracked; there are no large impact craters, but there are strange, light ridges in a scalloped pattern. The cracks are shallow; they are only 10 per cent darker than the surface they criss-cross, and give the impression of having been painted on.

The second painting shows Jupiter as it would be seen from Amalthea, a much smaller satellite orbiting only 113,000km (70,000 miles) above the cloud-tops of Jupiter. Amalthea was discovered in 1892, and is now known to have a reddish surface, with craters, ridges and troughs. The view would indeed be dramatic. It would seem to be an ideal site for an observatory; Jupiter's ever-changing atmospheric maelstrom would form an incredible, beautiful panorama for the astronomers in the observatory. Sadly, this is a scene which can never become reality, because the Pioneer and Voyager probes have found that Jupiter is surrounded by zones of radiation which would be quickly fatal to any astronaut incautious enough to enter them – and Amalthea moves in the main radiation zone.

Long-wave radiations had been detected from Earth in the 1950s, but the Pioneer and Voyager spacecraft also found short-wave radiation, caused by electrons moving very rapidly through a strong magnetic field, and it is these which rule out Amalthea as an observation point.

Jupiter's magnetic field is the strongest in the Solar System, and the radiation zones are at least 10,000 times more intense than the Van Allen zones surrounding the Earth. The magnetic axis is offset from the axis of rotation by 9.6°; the polarity is opposite to that of the Earth, so that, if it were possible to use a magnetic compass there, the needle would point south.

The Jovian magnetosphere is immense. On the night side of the planet the magnetotail may be as much as 650 million kilometres (400 million miles) long, so that at times it can even engulf Saturn.

THE VOLCANOES OF IO

Because Jupiter has so powerful a magnetic field, polar aurorae would be expected – and they do indeed exist; they were first imaged by the Voyager spacecraft in 1977 and were amply confirmed by the Galileo probe, which orbited Jupiter from 1995 to 2003. The aurorae form huge arcs on the night side of the planet. To a hypothetical Jovian observer with ultraviolet or infrared eyesight, they would be brilliant, though to us they would appear pinkish-red because of strong emissions in the hydrogen-alpha region of the electromagnetic spectrum.

There would also be other spectacular phenomena. Lightning on Jupiter is very intense; for example, in 1996 the Galileo probe recorded individual flashes hundreds of kilometres across, and no doubt there is appropriately deafening thunder as well.

Beyond the Jovian cloud-tops come the rings, unsuspected before the Voyager encounters. Jupiter's rings are dark, and quite unlike the glorious icy rings of Saturn; they seem to be produced from material released from the small inner satellites (including Amalthea) by the impacts of tiny meteoritic particles. There are three rings, now known as Halo, Main and Gossamer; they are tenuous, being to all intents and purposes unobservable from Earth. The Voyager and Galileo results show that the ring particles are very small; such particles have relatively short lifetimes in stable orbits, so the rings must be being continually replenished by material from the small satellites.

Io, the innermost of the Galilean satellites, is unique. It is very slightly larger than our Moon, and seems to have an iron-rich core overlaid by a mantle of partly molten rock; above this comes a sulphur-rich crust that is in a state of constant turmoil.

During the Voyager 1 pass, in March 1979, one of the JPL investigators, Linda Morabito, noticed on her computer screen 'an anomalous crescent' behind the limb of Io. It proved to be the gaseous plume of an active volcano (actually more analogous to a geyser). The plume was found to come from a heart-shaped feature now called Pele – a volcanic caldera, and the first erupting volcano ever observed on a body other than the Earth. Its bluish colour is due to the same 'Rayleigh scattering' that makes our own skies blue.

Many other volcanoes are now known on Io, active at different times, sending material upward to hundreds of kilometres; over 200 calderas have been identified. The surface itself is amazing: a variegated, pizza-like pattern of reds, yellows and whites, typical of the allotropes of sulphur, speckled with the black splotches of volcanoes.

The temperatures are high, and in some cases the temperature of the lava exceeds 1,000°C (1,800°F). The most powerful volcano, Loki, emits more heat than all the Earth's volcanoes combined. In February 2002 a particularly violent outburst close to Surt, site of a major eruption in 1979, was detected by the Keck telescope in Hawaii. It covered hundreds of square kilometres, and the temperature was over 1,200°C (2,200°F). The interior of Io is heated by tidal distortions caused not only by Jupiter but also by the other Galileans.

Since Io moves within Jupiter's lethal radiation zones, it seems safe to say that this moon is not likely to be visited by astronauts in the foreseeable future!

Polar aurorae on Jupiter, recorded several times by space probes and satellites, would appear reddish to human eyes, since there is none of the nitrogen and oxygen which give Earthly aurorae their green, blue or violet colouring, depending upon altitude. But they would be spectacular from above the cloud layers, as here. Io is seen as a yellow-orange moon, adding its glow to the scene. Even on Earth, the aurora is one of the most magical experiences in nature, second perhaps only to a total solar eclipse.

Overleaf:
An eruption on Io. The glow of fresh lava illuminates the base of a plume of gas on the horizon, and a lava-filled crack radiates from its source, behind typical mesa-like hills. In the middle distance is a dark lava lake, while in the foreground a now-extinct eruption emits a few last gasps.

EUROPA

The two inner Galileans, Io and Europa, are as different as they can possibly be. The reason can only lie in the fact that Europa orbits at a greater distance from Jupiter – on average 670,000km (416,000 miles), as against only 420,000km (262,000 miles) for Io – so the internal stresses affecting Europa are much less marked. There is absolutely no doubt that the surface layer is of water ice, and the idea of an ocean below the crust was suggested by two US astronomers, F. Reynolds and S. Squyres, as long ago as 1982, though no positive evidence was found before the data sent back by the Galileo spacecraft.

There seem to be 'icebergs' on the surface, and these may well shift around; water may force its way upward through cracks in the ice, freezing quickly. Over millions of years, this sort of activity has produced a complicated surface pattern, and few impact craters survive.

Can there be life in the sunless seas of Europa, assuming that (as seems overwhelmingly probable) those seas really exist? We cannot rule it out: life can appear in the most unlikely places. Within Europa, there could be volcanic action between the seas and the silicate mantle, giving rise to 'black smokers' similar to those on our own ocean beds, where we find the most alien lifeforms yet found on Earth; they depend not upon sunlight or oxygen, but on mineral-enriched super-hot water and on sulphur compounds. Europa may even provide us with a clue to the origin of life itself.

Facing page:
Suppose a meteorite could punch a temporary hole in a thin area in the crust of Europa, allowing in light from Jupiter. A scene like this might be revealed. A futuristic submarine probe is shown investigating, and alien life clusters around this oasis of life-giving heat and nutrient.

Below:
Seen from above Europa, the Conamara area shows colour variations from blue-green to brownish. The former areas may be dusted with water-ice, while the browner regions may contain dust and minerals. Jupiter's thin ring can just be seen, edge-on.

SATURN

Beyond Jupiter, moving at a mean distance of 1,425 million kilometres (886 million miles) from the Sun, lies Saturn – without doubt the most beautiful object in the Solar System. In composition it is not unlike Jupiter, but it is smaller and less dense. In fact, the planet's overall density is less than that of water; if Saturn could be dropped into a vast ocean, it would float! Its equatorial diameter is just under 120,000km (75,000 miles); it rotates quickly (in about 10.25 hours), and its globe, like that of Jupiter, is flattened. Saturn's cloud belts are much less prominent than those of Jupiter, and the planet's disc shows occasional bright white spots.

Saturn has been surveyed by several spacecraft. The most impressive results have come from the two Voyager probes in 1980 and 1981.

The glory of Saturn lies in its rings, unique in our experience. Once thought

We are floating high in the atmosphere of Saturn, above a latitude of about 15° north, looking east at the rising Sun. The painting gives some idea of the intricate structure of the rings, as revealed by Voyager; the many divisions are produced by the gravitational perturbations of the planet's moons.

to be solid or liquid, they are now known to be made up of icy particles. There are three main rings. The outer two, A and B, are separated by a 5,000km (3,000-mile) gap known as the Cassini Division in honour of its discoverer; closer in lies the semi-transparent Ring C, known also as the Crêpe or Dusky Ring. Still closer in there are particles making up what is termed Ring D, though it is not well defined. Outside the main system there are fainter rings – F, G and E, in order of distance. The whole system is much more complex than we used to think; the Voyagers showed hundreds of thin rings and minor divisions. The rings may be the remnants of a former icy satellite which was broken up, or they may be debris which never condensed into a satellite. They lie in the plane of Saturn's equator; extensive though they are, they seem to be not much more than a kilometre thick, so that, when seen edge-on, they appear as a thin line of light.

Titan, Saturn's only large satellite, remains an enigma. Apart from Ganymede, it is the largest satellite in the Solar System, and even in the 1950s it was known to have an atmosphere. The famous astronomical artist Chesley Bonestell visualized it as having a blue sky, with ice on the rocks; for the 1972 edition of *The Challenge of the Stars*, David Hardy did some research which showed that methane – known to be plentiful in the atmosphere – is greenish in quantity, and the sky was depicted accordingly. In the illustration, an expedition had just set off a charge of explosive to investigate the composition of Titan's surface and crust.

By the time of the 1978 edition, our ideas had changed; now Titan was believed to have a rocky core in which was bound water that contained dissolved ammonia. The satellite was now known to be covered by a layer of orange-red cloud, which is why Saturn is only dimly visible in the new painting. The same composition was kept for this version as in the 1972 one.

Right and above:
The changing appearance of Saturn's major moon, Titan, as our knowledge increases. In 1972 the sky was still usually portrayed as blue, as it had been since the 1940s (here it is green; see text). By 1978 Titan was seen as having an atmosphere of red smog, replenished by 'ice volcanoes'.

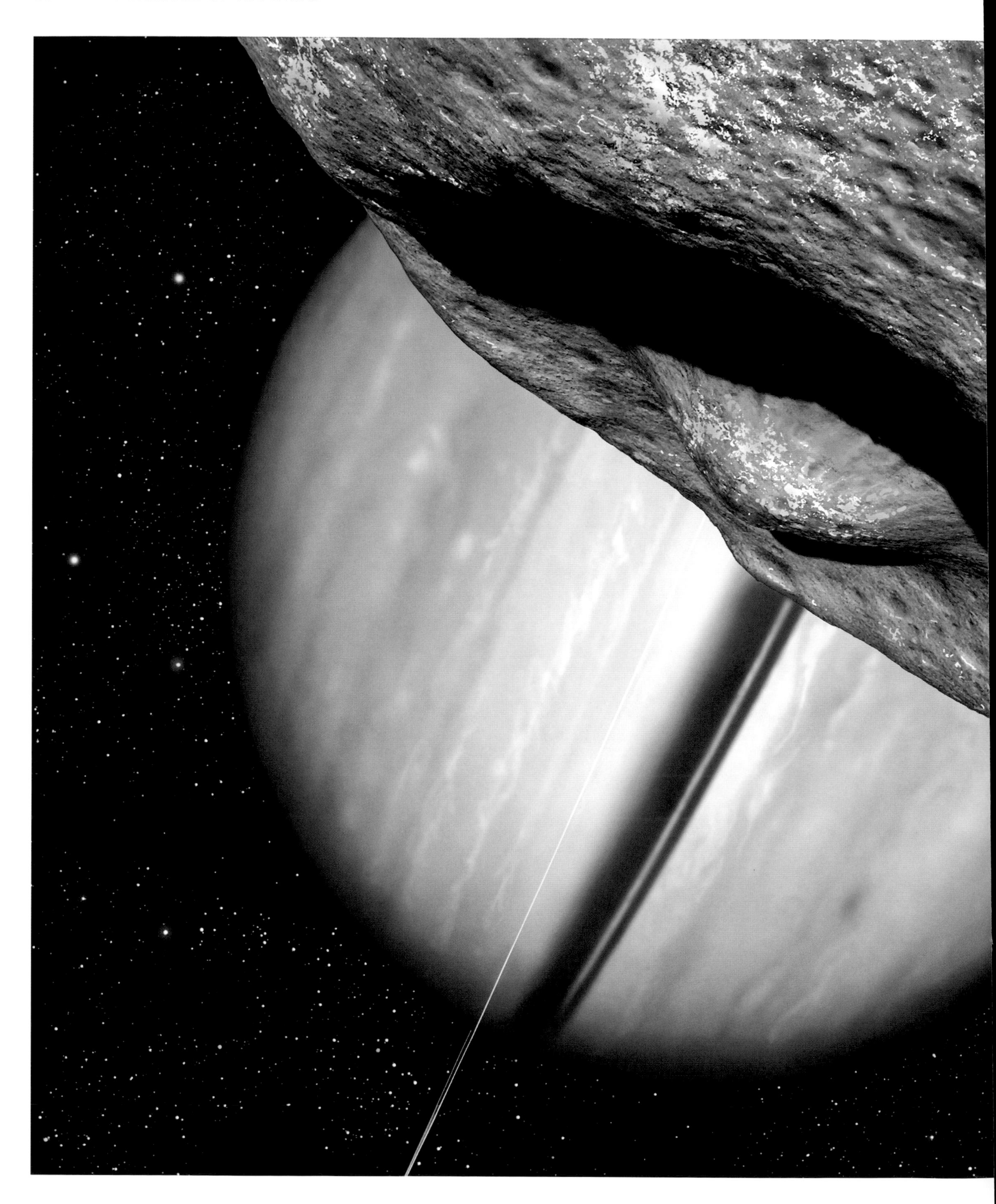

Saturn from its moon Mimas, with its giant crater Herschel. By contrast, Mimas's southern polar region is quite smooth. On the opposite side of the moon from Herschel are what seem to be fracture marks.

MIMAS, ENCELADUS AND THE SURFACE OF TITAN

Saturn has an extensive system of satellites, though only Titan is of planetary size. All the rest are much smaller; those with diameters of several hundreds of kilometres are Iapetus, Hyperion, Rhea, Dione, Tethys, Enceladus and Mimas.

In 1789 William Herschel discovered two of Saturn's inner satellites, Enceladus and Mimas. Both of these were surveyed from close range by the Voyagers in the early 1980s and found to be icy and cratered, though they are otherwise not alike.

Mimas orbits at a distance of 185,400km (115,300 miles) from Saturn, in a period of 22 hours 37 minutes. The globe seems to be made up of ice combined with some rock; the overall density is only slightly greater than that of water. The most remarkable surface feature is a large crater, fittingly named Herschel, which is 130km (80 miles) across – bear in mind that the diameter of Mimas itself is only 420km (260 miles)! Herschel crater was formed by an impacting body; had this been any larger, Mimas would have been shattered. The crater walls are 5km (3 miles) high, while parts of the floor are 10km (6 miles) deep; a central peak rises to a height of almost 6.5km (4 miles).

The gravity on Mimas is so weak that an explorer might have difficulty in distinguishing between 'up' and 'down' – hence the unusual angle chosen for this painting.

Mimas moves in the plane of the ring system, so an observer on the satellite would never see Saturn to best advantage. The rings would always be edgewise-on.

Enceladus, somewhat larger than Mimas, is very different. The surface is smooth and icy, with no crater as large as Herschel; there are areas almost crater-free, and evidently very young. There are now known to be powerful geysers erupting near the south pole, and there is a very tenuous atmosphere.

Titan is much the largest member of Saturn's retinue, and is in fact larger than the planet Mercury. There is a dense atmosphere, with a ground pressure 1.5 times that of the Earth's air at sea-level. On 14 January 2005 the Huygens probe, ferried by the Cassini spacecraft, made a controlled landing there, and sent back images and data for 72 minutes. The touch-down speed was less than 20 kilometres per hour, and Huygens came to rest on a thin crust, settling down some 20 centimetres below upon spongy, hydrocarbon material with about the consistency of wet sand. There were obvious drainage channels, and it seems that methane rain lands on the icy uplands and washes dark, organic material off the hills onto the plains. The hills themselves are of 'dirty water ice'.

Since the surface temperature is about –180°C (-300°F), life there seems to be unlikely. Following the Huygens landing, the whole of the globe was surveyed by the Cassini orbiter. No large liquid regions were found, but there were strong indications of methane lakes.

Titan is not welcoming, but it is certainly a world which can be reached. No doubt it will have plenty of surprises in store for the first astronauts.

OVERLEAF:
Beneath the clouds of Titan. A balloon probe, or dirigible, is a possible future method of exploring this fascinating world for months at a time. It would of course be rare to see Saturn. In the left foreground, dark organic material deposited at the top of a mountain is conducting heat and gases, and a small geyser can also be seen in the middle-distance. Although Cassini-Huygens has not shown such a terrain, Titan is a big world and this forecast has not been invalidated. Certainly ice-volcanoes seem likely.

IAPETUS

If Titan is the most important of Saturn's satellites, Iapetus is probably the most puzzling. It is almost 1,450km (900 miles) in diameter, and moves round Saturn at a distance of 3,540,000km (2,200,000 miles) in a period of 79 days. As it is the only relatively large Saturnian satellite to move away from the plane of the rings, it will be an ideal observation base for future space travellers. It is also well clear of Saturn's radiation belts, which are in any case much weaker than those of Jupiter.

The main peculiarity of Iapetus is that the two hemispheres are quite different. The leading hemisphere is as dark as a blackboard, while the trailing edge is bright and icy; this means that, as seen from Earth, Iapetus is very variable. It has captured rotation (it always keeps the same face towards Saturn), and when it lies to the west of Saturn the bright hemisphere is turned toward us, so Iapetus is visible with small telescopes at these times; when to the east of Saturn it becomes much fainter.

The demarcation line between the bright and the dark areas is not abrupt; there is a 240km (150-mile) 'transition zone'. Good images were sent back by the Voyagers, and the bright regions were seen to be cratered; little detail could be made out in the dark hemisphere.

Below:
Saturn seen from above Iapetus, as envisaged in 1972. It was then believed the brightness of one side of Iapetus was because of ice overlaid on dark rock. Here we see the border between these two surfaces. In the black sky are the stars of Scorpio, with the red supergiant star Antares (see page 80) just above the horizon.

Close-range images of Iapetus obtained from the Cassini spacecraft, from December 2004, produced some major surprises. In particular, it was found that there is a long ridge virtually coinciding with the geographical equator, which extends over 1300km and bisects the entire dark area. In places it rises to 30km, and is composed of mountains rivalling Olympus Mons on Mars – quite extraordinary on so small a world. It is not clear whether the ridge, often called the 'Great Wall', is due to a mountain belt which has folded upward, or is a surface crack through which material from below has erupted on to the surface. There is also a large, ancient impact basin delineated by steep, bright inner scarps, and one spectacular landslide was found.

It used to be thought that the dark material had been sent out from the interior, but it now seems more likely that it is a deposit which has come from space. There has been a suggestion that in the remote past Iapetus actually collided with the ring system.

Certainly Iapetus is unlike any other body in the Solar System. We have much to learn about the moons of Saturn.

Right:
It now seems that the dark material on Iapetus is organic. In this view the rings of Saturn are seen at a wide angle; Iapetus is the only satellite from which this is possible, though Saturn will appear only about as big as Earth does from the Moon – two degrees."

URANUS

Far beyond Saturn, at a mean distance of 2,867 million kilometres (1,783 million miles) from the Sun, moves the giant planet Uranus – just visible with the naked eye, but not identified as a planet until 1781, when it was discovered by William Herschel. It had been recorded before – John Flamsteed, the first Astronomer Royal, noted it six times between 1690 and 1715 – but it had always been mistaken for a star.

Uranus takes 84 years to complete one orbit around the Sun. Like all the giant planets, it has a short rotation period: 17 hours 14 minutes. The planet's equatorial diameter is 51,090km (31,770 miles). Ordinary telescopes show little of its rather greenish surface, and most of our detailed information comes from Voyager 2, the one probe to fly by the planet. On 24 January, 1986, Voyager flew past only 80,000km (50,000 miles) above the Uranian cloud-tops, and sent back excellent images before moving on to an encounter with the outermost giant, Neptune, in 1989.

Uranus has a gaseous surface. The upper atmosphere is made up chiefly of hydrogen, with a substantial amount of helium and about 2 per cent methane. Methane absorbs red light, which explains the greenish hue of Uranus.

In composition, Uranus differs markedly from Jupiter and Saturn. It seems to have little internal heat, and the rocky core is small; indeed, we cannot be sure there is a definite core at all. Uranus is made up largely of 'ices', not necessarily in solid form; probably the most plentiful ices are of water, methane and ammonia. It is more fitting to regard Uranus as an 'ice-giant' than as a 'gas-giant'.

The strangest thing about Uranus is the inclination of its axis. The tilt is 98° – and, since this is more than a right angle, the rotation is technically retrograde. This means the 'seasons' are most peculiar. First much of the northern hemisphere, then much of the southern hemisphere, is plunged into darkness for a period of 21 Earth years, with a corresponding 'midnight sun' in the opposite hemisphere, though for the remaining 42 Earth years of the Uranian year there is a more regular alternation of day and night. The rings and satellites lie basically in the plane of the equator, so the whole system 'rolls' along in its orbit rather than spinning upright like a top. As seen from Earth, there are times when a pole lies in the centre of the Uranian disc, with the equator lying round the edge.

Uranus. The double-spread painting for the original books depicted a scene from the innermost large moon, Ariel, with Miranda visible above the planet. Voyager showed the surface of Ariel to be covered by deep, criss-crossing valleys and canyons, so this painting proved predictive – though, the artist admits, this was mainly coincidence! Uranus is shown as a crescent; the horns extend from one side of the equator to the other, rather than from pole to pole.

The reason for this extreme axial tilt is not known - and there are other peculiarities as well. Uranus has a reasonably strong magnetic field (with polarity opposite to that of Earth), but the magnetic axis passes nowhere near the centre of the globe; it is offset by 58.6°. The planet has radiation zones similar in intensity to those of Saturn.

Uranus has a system of thin, dark rings, discovered in 1977 and since then imaged by the Voyagers and the Hubble Space Telescope. They seem to be made up mainly of particles a metre or two in diameter; their thickness is less than two kilometres (a mile or so). They are as black as coal, and it has been suggested that they may not even be permanent features of the Uranian system.

Five satellites were known before the Voyager 2 mission: Miranda, Ariel, Umbriel, Titania and Oberon. Other satellites have been found since, but all are below 150km (100 miles) in diameter.

The two largest satellites, Titania – 1,575km (980 miles) across – and Oberon – 1,513km (941 miles) across – are icy and crater-scarred; Titania, well imaged from Voyager 2, shows ice cliffs and trench-like features, while many of the craters on Oberon are dark-floored.

A photograph taken by Voyager 2 of one of the great surprises of its Uranus fly-by: the amazing ice-cliffs of the tiny moon Miranda, only 470km (290 miles) in diameter (see next pages). *(Photo courtesy NASA/JPL.)*

Ariel and Umbriel – diameters 1,158km and 1,169km (720 and 727 miles) respectively – are also icy and cratered, but they are not alike. On Ariel there are broad, branching, smooth-floored valleys which look as if they have been formed by liquids, but water is not a candidate, because of Ariel's small size and low temperature. Today Ariel is inert, but there is evidence of great tectonic activity in the past. Umbriel has a more subdued and darker surface, which gives every impression of being ancient.

Miranda, less than 500km (310 miles) across, has an amazingly varied landscape which presents real problems of interpretation; it is completely different from any other known planetary satellite.

Voyager 2 passed Miranda at a range of only 2,990km (1,860 miles), so the surface could be imaged in great detail. Naturally, only one hemisphere could be studied as the other was in darkness. At first sight Miranda appeared to be an excellent example of the catastrophic break-up and re-assembly of an icy body. There are several distinct types of terrain. Parts of the surface are cratered in the same way as the highlands of the Moon; there are brighter areas with cliffs and scarps, and ice cliffs towering to as much as 20km (12 miles). There are also three large, trapezoidal-shaped regions, known as coronae, covered with valleys, ridges and grooves. One of these, Arden Corona, has been nicknamed the Race Track!

It has been suggested that Miranda has been shattered and re-formed several times, and certainly the various types of terrain seem to have been created at different periods. But this would involve considerable heating, which in view of Miranda's small size and icy nature does not seem likely. The mysteries remain.

OVERLEAF:
Close to the south pole of Miranda, ice-cliffs rise up to 5km (3 miles) from the valley floor at an angle of 45-50°. This view may next be seen in the year 2029, as the Sun is shining on the pole of Uranus from below the horizon.

NEPTUNE

In size and mass, Neptune is similar to Uranus; it is slightly smaller, but considerably denser. Its mean distance from the Sun is 4,490 million kilometres (2,793 million miles), and the orbital period is 164.8 years, so that it has not yet completed one circuit since it was discovered in 1846. It was first identified by the German astronomers Johann Galle and Heinrich D'Arrest; its position had been predicted by the French mathematician Urbain Le Verrier from studies of the irregularities in the movements of Uranus. It is too faint to be seen with the naked eye, but binoculars show it as a starlike point and a small telescope reveals a bluish disc.

Uranus and Neptune may be near-twins, but, as with Venus and the Earth, they are non-identical twins. Neptune has an equatorial diameter of 50,500km (31,410 miles), as against 51,090km (31,770 miles) for Uranus, and the rotation period is shorter (15 hours 7 minutes), but it does not share Uranus's extreme axial tilt; the inclination of the axis of rotation is just over 28°, not very different from that of the Earth. The surface is much more active than that of the rather bland Uranus. There is a magnetic field, but, as with Uranus, the magnetic axis does not pass through the centre of the globe; it is displaced by 47°.

The two ice giants differ in another important respect. Neptune, unlike Uranus, has a strong internal heat source, so that the temperatures of the upper clouds are much the same on the two planets even though Neptune is so much further from the Sun.

As with Uranus, almost all our detailed information about Neptune has come from Voyager 2. There is an obscure ring system, as dark as that of Uranus and even less prominent; Voyager 2 surveyed it during its approach to the planet, and also discovered several small inner satellites. The most prominent feature on Neptune itself was the Great Dark Spot, which had a longer axis of about 10,000km (6,000 miles); its size, relative to Neptune, was about the same as that of the Great Red Spot relative to Jupiter. The probe saw other, smaller spots, as well as belts and high-altitude clouds; in some cases the clouds cast shadows on the cloud-deck beneath – a phenomenon never observed on Jupiter or Saturn, and certainly not on Uranus.

It was a great surprise when observations done with the Hubble Space Telescope in 1994 showed the Great Dark Spot had disappeared. In fact, the

Little was known about Neptune in 1972 other than that it was a bluish gas-giant planet with two satellites, Triton and Nereid, the former being large and the other very small. Triton's orbit takes it rather closer to the planet than the Moon is to Earth, at 354,760km (220,620 miles); the orbit is retrograde, and has a high inclination (160°), which gives it a view of both poles during its 5.75-Earth-day journey.

whole scene had changed – Neptune is indeed a dynamic world. Then, in 2003, the HST revealed that bands of clouds encircling Neptune's southern hemisphere were becoming progressively brighter, suggesting that the planet has seasons.

Before the Voyager 2 pass, only two satellites of Neptune were known, Triton and Nereid. Triton has a diameter of 3,700km (1,680 miles), and is therefore smaller than our Moon; it moves in a retrograde orbit some 350,000km (220,000 miles) from Neptune, with a period of nearly six days. There seems little doubt that it is a captured body rather than a bona fide satellite. Nereid is very small, and has a very eccentric orbit.

Nereid was not well imaged by Voyager 2, but the results from Triton were startling. The surface has a general coating of ice, but it was found that the southern pole was covered with nitrogen snow and ice, with some methane; Triton is the coldest world ever visited by a spacecraft. The greatest surprise was the discovery of active geysers. Apparently there is a layer of liquid nitrogen perhaps 30m (100ft) below the surface, where the pressure is high enough to keep it in the liquid state; if it migrates upward for any reason the pressure is relaxed, and the nitrogen explodes in a shower of ice and vapour, rising to heights of several kilometres before falling back. The outrush sweeps debris along with it, producing surface streaks.

There are no high mountains on Triton, but away from the nitrogen cap there are low-walled enclosures and complicated patterns of fissures. It is indeed strange to find activity upon a world so remote and so cold as Triton.

Nereid. This little moon can approach Neptune as close as 1.4 million kilometres (875,000 miles), as here, when the planet would look as large as Earth does from our Moon; but its very eccentric orbit takes it as far away as 10 million kilometres (6.25 million miles), from where Neptune would appear in the sky only half the size our Moon does from Earth. Nereid's orbital inclination of 29° allows us to see Neptune's rather faint ring-system at a wide angle. The moon is probably a captured ice asteroid.

OVERLEAF:
Geysers on Triton. Columns of gas rise almost vertically for several kilometres, then are sheared off by high-altitude winds into long streamers which travel horizontally for perhaps 150km (100 miles). In the painting, one such ice-volcano is erupting from inside a valley while another is just beyond the horizon. The outfall of dark material against the thin haze on Triton is exaggerated slightly.

PLUTO

Towards the end of the 19th century Percival Lowell, best remembered for his belief that Mars was inhabited, studied the movements of the outer giant planets Uranus and Neptune, and came to the conclusion that they were being perturbed by a yet more distant planet which had still to be discovered. He began a search for this hypothetical world from Flagstaff in Arizona, where he had established an important observatory and equipped it with a 24in (61cm) refracting telescope – then, as now, one of the best refractors in the world. Lowell died in 1916 without having had any success in this search, but in 1930 Clyde Tombaugh, working at the Lowell Observatory, discovered the planet very close to its predicted position. It was named Pluto, after the god of the Underworld. It is also probably no coincidence that the first two letters of Pluto are also the initials of Percival Lowell . . .

Since then, Pluto has set astronomers problem after problem. It does not seem to fit in with the general pattern of the Solar System. It is small: with a diameter of a mere 2,322km (1,444 miles) it is smaller than our Moon and Neptune's satellite Triton. It was precisely because of this smallness that Lowell

In 1972 it was unknown that Pluto had a moon. Not even the world's largest telescope would show the planet as a disc, and we were able to say only that it was a rocky body, smaller than Mars. A sea of liquid methane was envisioned, shown here from a cave (probably lava-tube) entrance. The Sun was shown as an intensely bright point of light, with no visible disc.

had overlooked it: it was far, far fainter than he had expected, and indeed a moderately large telescope is needed to show it at all. Second, its orbit is very eccentric. The distance from the Sun ranges between 4,366 million kilometres (2,715 million miles) at perihelion out to 7,370 million kilometres (4,583 million miles) at aphelion; Pluto's year is nearly 248 Earth years. The planet's orbit crosses that of Neptune – although, because it is inclined at an angle of 17°, there is no danger of a collision between Neptune and Pluto.

In 1978, J.W. Christy, from the US Naval Observatory at Flagstaff, discovered that Pluto has a satellite, now named Charon, after the boatman who ferried departed souls across the River Styx into the Underworld. Charon has a diameter of 1,200km (750 miles), more than half that of Pluto itself, and an orbital period of 6 days 9 hours. This is exactly the same as the rotation period of Pluto, so that, to an observer on the planet, Charon would remain fixed in the sky.

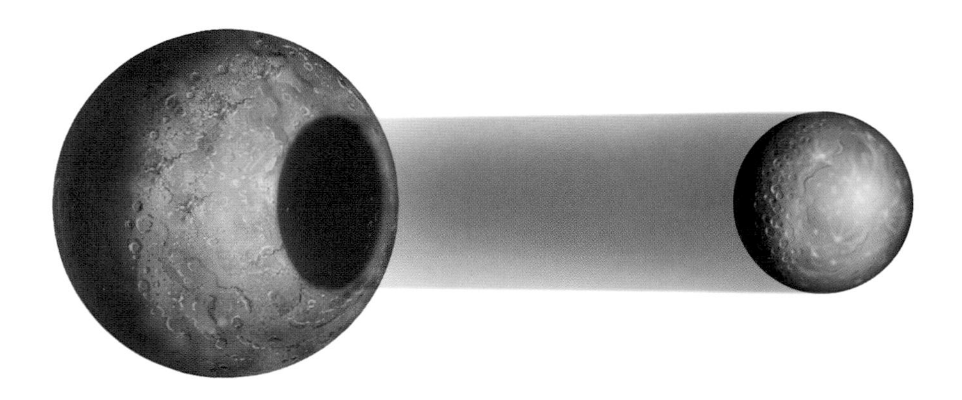

A diagram of Pluto and its moon, Charon, with their sizes (but not the distance between them) to scale. Charon can cast a large shadow on Pluto when it eclipses the Sun, but this occurs only every 124 years.

There seem to be distinct differences between Pluto and Charon. Charon has only one-seventh the mass of Pluto, and is grey while Pluto appears reddish; Pluto has greater reflecting power and contributes 80 per cent of the total light we receive from the system.

This situation is unique in the Solar System, and it may be reasonable to regard the Pluto/Charon pair as a double planet rather than a planet and a satellite. But is Pluto worthy of true planetary status? Serious doubts have arisen recently, following the discovery of many fairly large bodies moving in these remote parts of the Solar System; they make up the Kuiper Belt, named after the Dutch astronomer Gerard Kuiper, one of the first to suggest the belt's existence.

Many Kuiper Belt objects are now known, and at least one of them is larger than Pluto. Rather reluctantly, it seems that Pluto must now be regarded as nothing more than the brightest member of the swarm.

No spacecraft has yet been anywhere near Pluto, but the Hubble Space Telescope has shown a certain amount of detail on its tiny disc (never more than 0.11 of a second of arc in diameter). There seem to be a dark equatorial band and brighter poles; the axial inclination is 122° so that, as with Uranus, the rotation is technically retrograde. Charon moves in the plane of Pluto's equator.

Pluto has a tenuous but surprisingly extensive atmosphere, which may even extend as far as Charon; the distance between the surfaces of the two bodies is less than 20,000km (12,000 miles). However, the atmosphere may not always be present. Pluto passed through perihelion in 1989, and is now moving outward; when it is in the far part of its orbit, the intense cold may make the atmosphere freeze out. From Charon, Pluto would of course dominate the sky. There are about 14,000 Plutonian days in every Plutonian year, and this of course is also the number of orbits completed by Charon during that period.

Is there another large planet, well beyond the Kuiper Belt? There may be – but even if it exists, it will be so faint and so remote that it will be extremely difficult to locate.

OVERLEAF:
The temperature on the surface of Pluto can fall as low as -238°C (-396°F). One side of Pluto seems to be 30 per cent brighter than the other, and a bright highlight on Pluto suggests it may have a smooth, reflective area, such as the icy lake shown in the painting. Charon is bluer than Pluto, which appears slightly reddish, so the two are of a different composition. Pluto is the only body to rotate synchronously with the orbit of its satellite, so each always sees the same face of the other as they travel through space.

THE MILKY WAY

Our Galaxy, known as the Milky Way, is made up of around 100,000 million stars, of which the Sun is one. The Galaxy is about 100,000 light-years in diameter and is shaped like a double-convex lens – or, less picturesquely, like two fried eggs clapped together back to back. The Sun is about 26,000 light-years from the galactic centre; the whole system is rotating, with the Sun taking 225 million years to complete one orbit.

The main system is surrounded by the 'Galactic Halo', where there are many 'stray stars'. If we could stand on a planet of one of these stars, the scene might be very much as depicted here. Our Galaxy is actually a spiral, with indications of a 'bar' through its centre; from a range of 200,000 light-years it dominates the sky. The nucleus of the system consists mainly of cool, older, reddish stars (Population II), while the arms glow with hot, younger, bluish stars (Population I). There are stars of all types; some double, some variable, and many with – no doubt – planetary systems. There are also pinkish nebulae in which stars are being born; dark dust-lanes show up between the arms.

The Milky Way is a typical spiral galaxy, though slightly larger and more populous than average. In the far future it will collide with the Andromeda Galaxy, at present more than two million light-years away, and the spiral shapes of both will be destroyed. Probably the two systems will merge to form one giant elliptical galaxy (see pages 106–107).

When we on Earth look along the main diameter of the Galaxy, we see many stars in almost the same direction; this results in the appearance in our night skies of the pale band of light also known as the Milky Way. The stars there seem to be crowded but, as is so often the case, appearances are deceptive: the stars in the Milky Way are in general light-years apart.

The Milky Way, our Galaxy, seen from the hypothetical planet of a 'stray star' orbiting it. Our own Sun, a normal, yellowish star, is not of course visible here, but its position in the Sagittarius Arm is marked, near the top centre, by a white star (actually thousands of times closer). Above this is the Cygnus–Orion Arm; the outer arm, the Perseus Arm, is beyond the edge of the picture.

The 1972 painting of our closest stellar neighbour, Proxima Centauri, showed a featureless yellow-red surface on the rather feeble red dwarf star, which is only 1/150th as bright as our Sun. However, Proxima is a 'flare star', meaning its brightness can suddenly more than double. The Hubble Space Telescope has also detected dark 'starspots', like the Sun's sunspots. For this new version the original painting has been altered digitally in order to show spots and a large prominence – the harbinger of another flare.

PROXIMA CENTAURI

The nearest stars beyond the Sun are the three members of the Alpha Centauri system, which lie well south of the celestial equator. The two bright members of the system (Alpha Centauri A and B) appear as one star when you look at them with the naked eye, and outshine all other stars apart from Sirius and Canopus. In fact they are separated by a distance much the same as that between our Sun and Uranus; a small telescope shows the two stars well. Their orbital period is just under 80 years.

Proxima, the third member of the system, is very different, and at a range of

4.22 light-years it is actually the closest to us of all stars (A and B are 4.36 light-years from the Sun). In this scene A and B are behind us, but well visible is a familiar constellation: the W of Cassiopeia. It has an extra star to its left when seen from Proxima: the Sun, yellowish and fairly bright.

A habitable planet orbiting Proxima – i.e., one having liquid water – would need to have a year of only 10 to 12 Earth days. Shown here is a satellite silhouetted in transit.

Is Proxima a true member of the Alpha Centauri group? It is so far from A and B – about 13,000 Astronomical Units (one AU is the distance between the Earth and the Sun) – that it may not be gravitationally bound to them, and could eventually leave the system.

ANTARES

Above:
The 1972 book contained this painting of a red supergiant, Zeta Aurigae, which has a bright, blue companion and is an eclipsing binary. The blue star is about to be eclipsed, but will remain visible for some time shining through the tenuous outer envelope of gas surrounding the supergiant. On a hypothetical airless planet, 1,125 million kilometres (700 million miles) distant, the two stars cast bi-coloured shadows.

Few of the constellations look anything like the objects after which they are named. One exception is Scorpius, the Scorpion (often referred to, less accurately, as Scorpio); it is marked by a long line of bright stars, with patterns of stars making up the 'head' and the 'sting'. It lies well south of the celestial equator and from the latitude of the UK and northern USA it is inconveniently low down for observation; part of the 'sting' never rises at all. From southerly countries, the Scorpion is truly magnificent. Its brightest star, which does attain a respectable altitude when seen from the UK and USA, is Antares, 'the Rival of Mars'.

Antares is so named because of its redness; Ares was the War God of Greek mythology, equivalent to the Roman Mars. The star's colour makes it stand out at once; it is indeed the reddest of all the bright stars in our skies. With a visual magnitude of 0.96, it is the fifteenth most brilliant star in the sky. It is 600 light-years away, and 12,000 times as luminous as the Sun.

The most remarkable thing about Antares is its size. Its diameter is of the order of 400 million kilometres (250 million miles); if our Sun were as huge as this, even Mars would be engulfed. But Antares is of very low density, so its mass is no more than about ten times that of the Sun. It is enveloped in a cloud of gas which it has emitted, and which reflects its colour.

Antares is a typical red supergiant. Stars of this type were once thought to be young. We now know they are well advanced in their evolution, so have used up their main supply of nuclear 'fuel' and are drawing upon their reserves. Eventually disaster overtakes red supergiants; there is an abrupt collapse, followed by a tremendous explosion – a supernova. In the case of Antares, what will be left behind will be a very small, super-dense star made up of particles known as neutrons. When this will happen we cannot say, but happen it will.

Antares is not alone. It has a binary companion, moving round it in a period of nearly 900 years. This smaller star is slightly bluish in colour (the contrast with Antares making it appear more so), and its radiation causes the nebula surrounding Antares to fluoresce. From Earth the light of the companion is overwhelmed by the reddish glare of Antares, but even this smaller star is nearly fifty times as luminous as the Sun.

Can Antares be the centre of a planetary system? There is no definite reason why not, but a planet cool enough to support Earth-type life would have to be much further away from Antares than we are from the Sun. Neither will it have so extended a future; long before our Sun becomes powerful enough to destroy the Earth, Antares will have exploded.

Left:
The huge red star Antares is over 9,000 times more luminous than the Sun, but has a temperature of 'only' 3,065°C (5,550°F), compared with our Sun's 5,565°C (10,050°F). Even so, a planet would need to be 20 billion kilometres (12.5 billion miles) away in order to have an Earth-like atmosphere and liquid water. This idyllic landscape was deliberately painted to emulate the rather romantic work of the Hudson River School of artists, who painted the opening up of the USA's wild frontiers, such as Yellowstone Park and the Grand Canyon. The nebula around Antares is hidden by the atmosphere and clouds. *(From the private collection of David Egge.)*

ALGOL & FOMALHAUT

One of the most brilliant stars in the sky is Vega, in the constellation of Lyra (the Lyre). Near it is a much fainter star, Beta Lyrae. In 1784 a young astronomer named John Goodricke – who, incidentally, was deaf and dumb – noticed that Beta Lyrae did not shine steadily. It brightened and faded regularly, over a period of 12 days. We now know that Beta Lyrae is made up of two stars, almost in touch, moving around their common centre of gravity. When one component passes in front of the other, the total light we receive from the system drops.

With Beta Lyrae the two components are not so very unequal, so that the light-curve is complicated and the variation subtle. Not so, however, with the most famous of these 'eclipsing binaries', as such star systems are called: Algol, in Perseus, known to the old Arabs as the Demon Star, though there is no evidence that they realized there was anything unusual about it. For most of the time Algol shines as a star of magnitude 2, about equal to Polaris, but every two and a half days it fades, taking four hours to fall to below magnitude 3, remaining at minimum for a mere twenty minutes before brightening once more. Of the two components, the brighter (A) is 105 times as luminous as the Sun; the fainter (B) has only three times the Sun's luminosity. The main minimum occurs when B passes in front of A; when B is hidden by A, the drop in magnitude is too small to be noticeable. Eclipsing binaries are not uncommon, but few are bright enough to be seen, like Algol, without optical aid.

Fomalhaut, in the little constellation of Piscis Australis (the Southern Fish), is a bright star of magnitude 1.2. One of our nearer neighbours, at a distance of 22 light-years, it is twelve times as luminous as the Sun. It is of special interest because it is now known to be surrounded by a cloud of material which may well indicate the presence of at least one planet.

Over one hundred stars have been found to be the centres of planetary systems; no 'extrasolar planet' has been seen directly through a telescope, but they can be detected by indirect methods. Most of them are large – gas giants more similar to Jupiter than to the Earth – and orbit their parent stars closely, but, judging from the characteristics of the cloud around Fomalhaut, a planet there may be much more distant – perhaps as far out as Neptune is from the Sun.

Fomalhaut is the most southerly of the first-magnitude stars visible from Europe. It is easy to locate because it lies in a very barren region of the sky; the Southern Fish contains no other star above the fourth magnitude.

Below:
The binary star system Beta Lyrae, as imagined in 1972. When the fainter star passes in front of its companion, Beta Lyra dims as seen through a telescope. For many years it was thought that the pair of stars emitted an expanding spiral of red-glowing hydrogen, as here, but a more modern depiction would show one of the stars surrounded by a disc of hydrogen which it is drawing from the other star, both being enveloped in a shell of gas. (See the illustration opposite.)

Right:
An Algol-type binary. The small blue-white star is surrounded by a spiral of gas drawn from the orange giant. Though a similar system to Algol, this is actually V471 Tauri, in which the mass-gaining star is a white dwarf; in Algol it is a main sequence B8 star about three times as big as the Sun, while the mass-loser is a yellow K2 star three and a half times as big. *(From the private collection of Dirk Terrell.)*

Overleaf:
In October 2002 a 'warped disc' was found around the brilliant white star Fomalhaut. Within this is a Saturn-sized planet (shown here as also *looking* rather like Saturn, though of course we do not know if it is ringed). This planet and two inner ones leave trails in the cold, doughnut-like dust-cloud. Fomalhaut is only about 200 million years old, compared to our Sun's 4.5 billion years. *(Courtesy of PPARC.)*

TAU GRUIS

No telescope yet built is able to show planets of other stars: a planet is much smaller than a normal star, and has no light of its own. However, other methods of detection are available. A massive planet orbiting a fairly small star will make the star 'wobble' very slowly and very slightly. The first planet to be identified through this 'wobble' effect orbits the star 51 Pegasi; it was detected in 1995. Many others have followed; the hundredth, announced in September 2002, is associated with the obscure star Tau Gruis, in the southern constellation of the Crane. The star is 100 light-years away, and is just over five times as luminous as the Sun.

What makes this planet so interesting is that, although it seems to be a giant, rather more massive than Jupiter, it lies only about 370 million kilometres (230 million miles) from the star – equal to the distance between our Sun and the main asteroid belt; its orbital period is 3.5 Earth years. This means its orbit lies nearly within the so-called habitable zone round Tau Gruis – that is to say, the zone which is at a tolerable temperature for life in theory to exist, provided other (extremely important!) requirements are met. In the Solar System, Venus is on the inner limit of the habitable zone and Mars at the outer limit, with the Earth in the middle. Since Tau Gruis is brighter than the Sun, its habitable zone lies further out. The detected planet is almost certainly a gas giant, not suited to advanced life, but it might well have satellites which would be good candidates; the system could be like that of Jupiter, where there are four satellites (the Galileans) of planetary size. The satellites – if they exist! – could be Earth-like worlds, with oceans and oxygen-rich atmospheres.

Giant planets are, of course, the easiest to detect over interstellar distances. However, space-based efforts are currently being made to discover Earth-like planets of other stars. Of course, the ideal would be to observe extrasolar planets telescopically, and it is possible the James Webb Space Telescope, due to be launched before the end of the present decade, may be equal to the task. There can be little doubt that the Universe holds many planets suitable for life, but whether life will appear whenever the environment is tolerable is something we do not yet know.

At least it is *possible* there are sentient beings in the system of Tau Gruis. In their sky our Sun would be very obscure; to detect even Jupiter from a range of 100 light-years would require very powerful instruments.

Facing page:
The first exoplanets were discovered in 1995, and the hundredth was announced in September 2002. This was a Jupiter-like planet orbiting the star Tau Gruis, 100 light-years from Earth, in the southern hemisphere's constellation Grus (the Crane). The planet is about as far from Tau Gruis as the asteroid belt is from the Sun in our Solar System – 2.5 Astronomical Units. *(Courtesy of PPARC.)*

Below:
Produced in December 2001, this is a depiction of an Earth-like exoplanet as seen from a small, close satellite above which its sun is rising. The requirements for life are very complex, and we have as yet no evidence of life beyond our own planet. The discovery of even micro-organisms on Mars or Europa would lend credence to the widespread idea that such life *can* exist, and this is one of the main drives for such research – and for space travel. *(Courtesy of PPARC.)*

BROWN DWARFS

The dividing line between a 'hot Jupiter' and a forming star is very slim. Objects known as 'brown dwarfs' have long been suspected, and in 1995 observations with the Hubble Space Telescope confirmed the existence of one such dwarf, Gliese 229B. This is the small companion of the cool red star Gliese 229, 19 light-years away from us in the constellation of Lepus. Gliese 229B is thought to be between 20 and 50 times as massive as Jupiter – too massive and hot to be classed as a planet, but too small and too cool to shine like a star. It is at least 100,000 times fainter than the Sun, and is the dimmest object ever directly observed orbiting another star.

Overleaf:
What a brown dwarf might look like. This one is the companion to a bright white star; the foreground planet is close enough to the star to possess bubbling lava and fire-fountains. High-altitude clouds on the day side of the brown dwarf appear light and almost violet, in contrast to the night side, where they appear dark against the internal red glow.

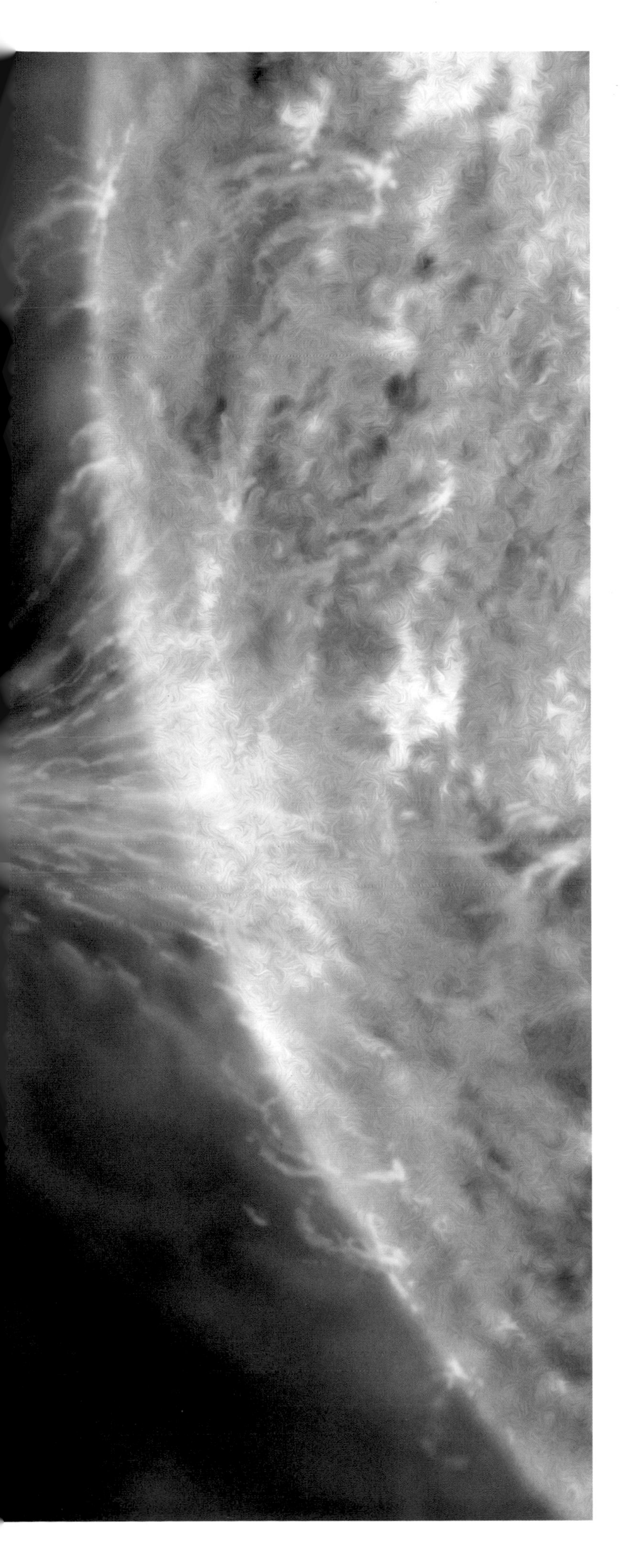

BLACK HOLES

A black hole is one possible consequence of a supernova outburst. A very massive star exhausts its nuclear 'fuel' and collapses with catastrophic suddenness; the collapse cannot be stopped, and so the star becomes smaller and smaller, denser and denser. As it does so, the escape velocity rises. A less massive star will collapse to form a neutron star, in which the matter is packed incredibly densely; this is what will happen to Antares (see page 80). When a more massive star collapses, however, eventually the escape velocity reaches 300,000km (186,000 miles) per second – the velocity of light. The collapsed star has surrounded itself with a 'forbidden zone' from which nothing can escape.

Obviously we cannot see a black hole, because it emits no light or radiation at all (although the disc of matter that surrounds it does), and so we can detect it only by its effects upon bodies we *can* see. One of the first suspected examples of a black hole was Cygnus X-1. In 1972 a blue supergiant star in Cygnus, about 30 times as massive as the Sun and 6,500 light-years away, was found to have an invisible companion which emitted X-rays; the companion was assumed to be either a neutron star or a black hole. Doppler studies showed that the blue star has an orbital period of 5.6 days, and from this it can be calculated that it is 14 times as massive as the Sun – too massive to be a neutron star. A black hole is the most likely explanation. It is pulling material away from its companion, and this material is strongly heated before disappearing into the black hole; hence the X-radiation. Excess gas is ejected from the disc of infalling matter, and forms two very active jets above the rotational poles of the black hole. The view from afar would indeed be striking.

Many other cases have since been found, and the reality of black holes no longer seems in doubt. Thankfully, our Sun can neither explode as a supernova nor collapse to form a black hole; it is not nearly massive enough.

A black hole. Matter drawn from the blue giant star would become ionized and be greatly accelerated, forming X-rays; as gas spirals into the accretion disc, excess material shoots out at incredible speed in jets above and below the black hole.

ETA CARINAE

It is commonly said that no telescope can show a star as anything but a point of light. The Hubble Space Telescope has shown this adage to be false, at least in a few cases. One of its most spectacular images shows Eta Carinae, a star (and associated nebulosity) in the southern constellation of the Keel, close to the Southern Cross. At present Eta Carinae is barely visible with the naked eye, but it has had a chequered history, and for a while during the 19th century it was the second brightest star in the entire sky – inferior only to Sirius.

It was recorded by Edmond Halley in 1677 as being of the fourth magnitude. It remained between magnitudes 4 and 2 until 1827, when it blazed up to the first magnitude. By 1837 it had become as bright as Alpha Centauri, and it reached its peak in 1843 at magnitude –0.8, outranking Canopus. A decline then set in; by 1870 the magnitude had fallen to 6, and ever since then the star has hovered on the brink of naked-eye visibility.

At maximum the luminosity was about six million times that of the Sun, and in fact this has not really declined a great deal even though the visual magnitude has fallen: most of the emission is now in the infrared region of the electromagnetic spectrum. The star is associated with nebulosity, and superimposed upon this is a dark, dusty region known (because of its shape) as the Keyhole Nebula.

What seems to have happened is that, in the years following 1834, a massive explosion threw off a shell of material from the surface of the star. As the shell expanded, the star seemed to brighten; after 1843 the shell cooled, dimmed and finally became opaque, blotting out the light of the star within. We now see the shell as a tiny nebula, nicknamed the Homunculus Nebula (again from its shape). When seen through a telescope Eta Carinae looks quite unlike an ordinary star. It is orange in colour.

Eta Carinae seems to be an ideal candidate to explode as a supernova in the relatively near future; it certainly has a limited lifespan. When it explodes, which it may do at any time, its likely fate is to create a black hole – the star is thought to be exceptionally massive.

The Hubble Space Telescope image shows a great amount of detail, but there are still doubts about the object's true nature. It has been suggested we are dealing with not one star but two – a close binary. However, it is more generally believed that Eta Carinae is a single super-massive star. Observers keep a careful watch on it; a new outburst is quite possible in view of the fact that the star is so unstable.

It does not seem very likely that Eta Carinae has a planetary system, though, like any star, it could possess 'captured' planets; if it has, the chances of life thereon are very remote. Even an Earth-sized planet, orbiting at a respectful distance, would have a most uncomfortable climate, and living organisms would also have to endure the flood of infrared radiation pouring out all the time.

If Eta Carinae does go supernova, it will for a brief period be truly magnificent in our skies, even though 6,000 light-years away. At its peak it was one of the most luminous stars in the entire Galaxy, and even now we know of few to match it. In our experience, Eta Carinae is in a class of its own.

Facing page:
Hubble images of Eta Carinae show, in amazing detail, two 'balloons' of gas and dust expanding from the poles of the star and reddening its light, plus a thin equatorial disc, all travelling outward at 650km (400 miles) per second. In 100,000 years' time Eta Carinae will probably end its life as a supernova and then a neutron star. A suitably placed planet might show a scene like this, with the nebulae in perspective and the equatorial disc in clear view.

ABOVE:
The 1972 book contained this illustration of the aftermath of the explosion of a star into a nova, increasing its light by 70,000 times and vaporizing all of its inner planets. We are observing from the 'Pluto' of its system. The sky is enriched by an aurora-like display caused by the outer layers of the star as they expand and fluoresce in the ultraviolet radiation. The star is a binary, but the red giant companion looks ghostly behind the bright white dwarf which has been robbing it of its mass. The bright star is rapidly fading, but may explode again, perhaps several times.

FACING PAGE:
Prequel. The painting on the opposite pages has been manipulated digitally to show the scene immediately after the nova explosion. Tall mountains have been melted down like candle-wax. On the day side of the planet, volcanoes erupt, lava flows from numerous fissures, and a cloud of super-heated steam blasts upward as an ocean is vaporized. The surface near us is cracked and scattered with glass-like globules of rock.

OVERLEAF:
If planets orbiting pulsars really exist, they must be very strange objects. When in its most active phase, the radiation from the pulsar would render the surface red-hot, though later it would become very cold as the pulsar 'spun down'. The sky is filled with shells of fluorescing gas, caught in whirling magnetic fields – a magnificent and beautiful sight, but one which can never be seen by human eyes, except in the imagination.

NOVAE AND PULSARS

'Nova' means 'new', but a nova is not a new star at all. What happens is that a formerly obscure star flares up to many times its normal brilliance, remaining bright for a period before fading back to its former state.

A typical nova involves a binary system. One component is a normal star, the other a white dwarf – a very small, very dense and very highly evolved star. The white dwarf pulls material away from its companion onto itself, and finally accumulates so much excess matter that it becomes unstable. There is a brief outburst. When this has subsided the star returns to its normal state. Some stars have been seen to suffer more than one outburst; these are known as recurrent novae (T Pyxidis and T Coronae are good examples).

A nova throws off a shell of gas, which expands rapidly. It was once thought that the expanding shell must be symmetrical and smooth, but the Hubble Space Telescope has shown that some novae, such as T Pyxidis, produce thousands of gaseous 'blobs', each about the size of our Solar System, within an area one light-year in diameter. Note that a nova is quite different from a supernova, which involves the virtual destruction of a star; a star that has gone supernova can never explode again, and will become either a neutron star or a black hole.

In 1054 a supernova appeared in Taurus (the Bull). It became bright enough to be visible with the naked eye in broad daylight. Nearly two years passed before it dropped below naked-eye visibility. That great event left behind the Crab Nebula, discovered in 1758 (and at first thought to be a comet). It is 6,000 light-years away, and contains a pulsar – a rapidly-rotating neutron star, only a few kilometres in diameter, but as massive as the Sun. A very intense magnetic field concentrates the radiation emitted into two cones, coming from the magnetic poles. Those magnetic poles do not coincide with the poles of rotation, and so, as the Crab pulsar rotates, 30 times per second, it sweeps a beam of radiation across our line of sight – the effect may be likened to that of a watcher on the sea-shore being regularly illuminated by the rotating beam of a lighthouse.

As a pulsar spins and emits energy, its rate of rotation slows by a measurable amount. The period of the first pulsar to be discovered (by Jocelyn Bell Burnell, in 1967) is lengthening by a thousand-millionth of a second each month, so that in 3,000 years' time the period between pulses will be 1.3374 seconds instead of the current 1.3373 seconds.

Obviously there must be many pulsars whose beams do not sweep over the Earth, and which therefore we cannot detect.

In 1992 astronomers were puzzled by the reported discovery of three planets orbiting the pulsar PSR 1257 + 12. If these reports were accurate, and if indeed the planets originally orbited a star which exploded as a supernova, they could be the cores of old gas giants which have moved closer to their primary; even the furthest is as close to the pulsar as Mercury is to our Sun. Another possibility is that they might have been formed more recently from the debris left in the wake of the supernova.

GLOBULAR CLUSTERS

Star clusters are of two kinds. Open or galactic clusters are merely groups of stars – a few dozen to a few hundred – which share common motion in space, and presumably have a common origin. Globular clusters are vast symmetrical systems containing from 10,000 to at least a million stars, and are so condensed that near their centres the stars may be only light months apart. They may be regarded as galactic satellites; our own Galaxy has at least two hundred of them orbiting around it.

Globular clusters are very old, so their leading stars are well advanced in their evolution, tending to be yellow or orange. Because they are thousands of light-years away, few globular clusters are visible to the naked eye; the brightest are Omega Centauri and 47 Tucanae in the southern hemisphere and the Hercules cluster in the northern.

Until recently it was thought that planets would be very uncommon in globular clusters, but the discovery in 2003 of a planet orbiting an old star in the cluster M4 seems to show this is not the case. Planets are one thing, *inhabited* planets quite another, of course. If inhabited planets existed in a globular cluster, the beings there would enjoy fascinating skies. Many stars would be bright enough to cast shadows, and the nights would know no proper darkness.

From a planet in a nearby area of the Galaxy, a globular cluster would dominate the entire sky.

Above:
Perhaps one of the most exotic illustrations in the 1972 book was this attempt to show a really alien form of life, in this case inhabiting a planet of a star inside a globular cluster. There can be no night here, since the sky is always filled with thousands of stars, far closer than those of our own skies, and brilliant. Hardy postulated oxygen-filled sacs floating on an ocean and forming 'rafts'; when these mature they become free-floating, carried by the winds of a carbon-dioxide atmosphere.

Facing page:
The brightest globular cluster, as seen from Earth, is Omega Centauri (NGC5139), which to us appears near the plane of the Galaxy, halfway in from the Sun towards the centre. Here we see the view from a hypothetical inhabited planet in that part of the Galaxy, but 300 light-years away from the cluster, which is currently at the zenith.

THE DEATH OF THE SUN

Our Sun is now a stable, well behaved star, but its supply of nuclear 'fuel' will not last indefinitely. Over the next billion years it will become slowly more luminous, and the Earth will become intolerably hot, from our point of view. Worse will follow. Four billion years hence the Sun will have become so powerful an energy source that the surface temperature of the Earth will reach 190°C (374°F), and the oceans will evaporate. Another billion years, and the Sun will become a giant star; Mercury and Venus will be swallowed up. The Sun will have lost some mass during these processes, and its decreased gravity will allow the planets to move outwards, but even so the Earth cannot escape its doom: the planet's atmosphere will be driven away, and the surface will become barren cosmic slag, perhaps emitting a few last gasps as any remaining gas is driven off from a cracked desert of lava.

The Sun will not remain a giant; it will throw off its outer layers to produce what is termed a planetary nebula, and what is left will shrink to become a very small, very dense star – a white dwarf – still shining feebly because of its continued contraction. After an immensely long period, all light and heat will leave it, and the Sun will become a cold, dead globe – a black dwarf – perhaps still circled by the ghosts of its remaining planets.

It is the destiny of any solar-type star to end its career as a black dwarf, but the process is so leisurely that the Universe may not yet be old enough for any black dwarfs to have formed.

Overleaf:
The planet of a red giant. This world still has a close satellite (our own Moon will have receded by this period in the Sun's life, if it still exists), and some atmosphere remains – though of course no life. The surface is cracked lava, still bubbling and degassing in places on the day side of the planet.

NEBULAE

Facing page:
A spectacular view of the Orion Nebula would be obtained from a planetary system a mere 40 light-years away. Often the colours ascribed to celestial objects in photographs seem to be quite arbitrary. The human eye is most sensitive to green wavelengths. This painting is a compromise between a photograph and what would actually be seen from such a well placed planetary system.

Gaseous nebulae are clouds of gas and dust in the space between the stars. Inside them, new stars are being born. Some gaseous nebulae shine only through being illuminated by stars in or close to them; these are reflection nebulae. Others – emission nebulae – are illuminated by short-wave radiation from very hot stars, and emit a certain amount of light on their own account.

The most famous of all nebulae lies in the constellation of Orion, making up the Hunter's Sword. It is officially listed as M42, indicating that it was the 42nd object in the catalogue of clusters and nebulae drawn up by the French astronomer Charles Messier in 1781. Ironically, Messier was not interested in nebulae; he was searching for comets, and merely listed clusters and nebulae so that he could avoid confusing them with comets! M42 is easily visible with the naked eye, and even a small telescope shows it well, but this bright region is only part of a huge nebula which covers most of the constellation. The nebula is 1,500 light-years away, and, although very large, is very rarefied; it has been said that, if you could take a 2.5cm (1in) core sample right through it – a total distance of 30 light-years – the total weight of materials collected would just about balance a pound coin.

Photographs taken with modern instruments, such as those in the Hubble Space Telescope, give glorious views of the Orion Nebula. However, these vivid colours would not be seen with the naked eye, which does not have sufficient 'light-catching' power; to detect them it is necessary to use photographic methods, powerful telescopes, long exposures, sensetive emulsions and electronic devices such as CCDs. The colours are genuine enough, but they are usually enhanced for public consumption, or simply to highlight specific wavelengths, such as red for the spectral hydrogen–alpha line.

The stellar birthplaces inside the Orion Nebula are made up of hydrogen; there are young, hot stars and fuzzy blobs known as 'proplyds', which seem to be embryonic planetary systems. Near some of the bright stars in the nebula, we see shock-waves – i.e., fronts where fast-moving material is hitting slow-moving gas.

The Orion Nebula shines because of the very hot, massive stars that make up the 'Trapezium', Theta Orionis, which is easy to see with a small telescope; the four main stars are indeed arranged in the shape of a trapezium. However, the dust means we cannot peer into the heart of the nebula, where there are stars we will never see at visible wavelengths because their lifetimes are too short for them to 'burn a passage' into outer space. One such is the Becklin–Neugebauer Object (BN), deep inside the nebula, a young, immensely powerful star which we can detect only because of its infrared radiation, which can slice through the intervening dust.

Most nebulae such as M42 are formed from the debris of supernovae. Since stars such as the Sun are formed inside nebulae, every one of us can proudly claim to be made up of the atoms from a supernova explosion. As Joni Mitchell wrote in her song 'Woodstock', 'We are star-dust.'

Many star-forming nebulae are known – such as the Eagle Nebula in Serpens (the Serpent), featured in a magnificent photograph from the Hubble Space Telescope; there are three columns of gas and dust, each more than a light-year long, inside which stars are forming. Near the tops of the columns, the radiation from the fledgling stars is evaporating the gas, so that eventually it will be dissipated; these structures are known as Evaporating Gaseous Globules (EGGs). Another interesting nebula is the North American Nebula in Cygnus. Binoculars show it really does recall the outline of the North American continent, with darker material in the position of the Gulf of Mexico.

JETTING GALAXIES

Our Galaxy, with its hundred billion stars, is by no means exceptional, though slightly above the average in size and mass. It measures 100,000 light-years across; the centre of the system lies about 26,000 light-years from us. We cannot see through to that centre, beyond the lovely star clouds in Sagittarius, but we believe the nucleus contains a black hole, Sagittarius A-* (pronounced Sagittarius A-star), with a mass at least two million times that of the Sun.

It is now generally believed that a typical galaxy has at its core a massive black hole – up to tens of millions of times as massive as the Sun. This sweeps up interstellar material and swirls it into an accretion disc, which persists for some time before the hot gases finally plunge over the 'event horizon' into the black hole. The thick disc of rotating matter tends to prevent the radiation generated by the hole from escaping at the disc's sides; this radiation is therefore blasted into space in narrow jets above and below the disc, travelling at speeds close to that of light. If one of these jets points directly towards the Earth, the intensely radiating galaxy is sometimes known as a 'blazar'. Doppler measurements made by the Hubble Space Telescope show that stars and gas on one side of a galactic core travel towards us at up to 400km (250 miles) per second, while on the other side they travel away from us at the same speed. Needless to say, any observer would need to be at a safe distance from those jets!

A 'jetting galaxy', as seen from a world located in a satellite galaxy. This world is actually a moon; the crescent Earth-like planet it orbits is providing most of the illumination for this scene, their parent star being below the horizon.

COLLIDING GALAXIES

We know that the Universe is expanding, and it is often said that all the galaxies are receding from each other. This is not strictly correct; galaxies tend to form groups, and each *group* of galaxies is receding from each other *group*. The speed of recession increases with distance. We can now observe galaxies over 12,000 billion light-years away receding at more than 90 per cent of the velocity of light.

Our Galaxy is a member of what is termed the Local Group, which contains three large spirals and over twenty smaller systems. The two senior members are our Galaxy and the Andromeda Spiral, M31, well over two million light-years away; it is larger and more massive than our Galaxy.

Collisions between galaxies within groups are not uncommon. In general the individual stars do not hit each other, though the material between them is colliding all through the encounter. The situation may be likened to two orderly crowds passing through each other as they move in opposite directions. In about three billion years our Galaxy will collide with M31 and the spiral forms of both will be lost; the two galaxies will merge to make up one colossal elliptical system.

We do not know whether human beings will have learned how to cross interstellar space by then – or if indeed we ever will. All kinds of interstellar spacecraft have been proposed, but the problems are tremendous. The limitation is of course the speed of light, which according to Einstein we can never equal. We have sent people to the Moon and spacecraft out to the planets, but for now we have to admit that, without great advances in technology, we cannot reach out to the stars.

Facing page:
Collisions between galaxies are quite common, despite the immensity of space; our own Milky Way will collide with its neighbour, M31 in Andromeda, in about three billion years' time. The result of collisions between galaxies is usually an elliptical galaxy, often with a strange 'tail' that eventually disappears. Here two galaxies are just beginning to pass through each other. The icy foreground planet has a civilization – its architecture perhaps influenced by the constant sight of these mighty spirals in its sky.

Below:
Several drives for interstellar flight were explored in the 1972 book. One that made it through to the second edition was the photon drive. Effectively, this emits a beam of light from the giant parabolic reflector (here cooling upon arrival) that propels the ship – rather in the way that the 'primitive' ion beam motor does for the lunar probe back on pages 14–15. The background here is the beautiful Trifid Nebula, which, like the Orion Nebula, is the birthplace of stars. The lightship has arrived at an Earth-like world, with its moon.

INTERSTELLAR TRAVEL AND COMMUNICATION

Above:
This 1972 illustration shows a radio telescope on the planet of another star, with a huge moon in its sky. The planet's inhabitants may be trying desperately to contact us!

Facing page:
A starship – its propulsion system has yet to be designed – arriving in an alien solar system. The planet is obviously Earth-like, but is it inhabited? And, if so, what sort of reception will the crew receive? Only time will tell . . .

Are we alone in the Universe? The answer must surely be 'no'. Everything we know indicates that the Earth is an ordinary planet moving round an ordinary star, and there is no reason to suppose it is unique in supporting intelligent life. However, it is true that so far we have no proof of life on another world.

Direct contact would involve technology that is currently beyond our comprehension. Rockets will not suffice – they are much too slow. Science-fiction writers have proposed devices of all kinds, such as asteroid arks or generation ships in which only the descendants of the original crew arrive at the destination. The immense distances would have to be overcome somehow, but – sadly – space warps, time warps, teleportation and thought-travel are beyond our present abilities.

Communication with other civilizations is a different matter, and here the outlook is much more promising. Radio waves travel at the same speed as light, and if an advanced race existed within reasonable range we could well get in touch. Indeed, the search has been going on for some years, under the guidance of organizations such as SETI (the Search for Extra-Terrestrial Intelligence).

Any message, to be universally comprehensible, must be mathematical; after all, we did not invent mathematics, we simply discovered it, and other civilizations will have made the same discovery. Originally a wavelength of 21.2 centimetres was chosen as a likely one on which radio broadcasts might be made; this is the wavelength of radio emissions from the clouds of cold hydrogen spread widely throughout the Galaxy, and thus it is reasonable to assume it is a wavelength to which alien radio astronomers might pay close attention. More recently, other wavelengths have been proposed as equally compelling.

In the past few years, *SETI@home* has been organized. This allows anyone who has a PC (Mac or Windows) to take part in the search for extraterrestrial civilizations by downloading a screen-saver which, when a computer is idle, will download 300k of data to be analysed for possible ET signals. Up to now the results have been negative, but it is a measure of our changed attitude that the search has been regarded as worth undertaking at all.

The Space Age has been in progress for less than half a century, but we have made amazing strides since the flight of Russia's tiny Sputnik 1 ushered in the new era. Given full international cooperation, and freedom from the wars that have plagued the human species throughout its history, there is no limit to what we can achieve. We cannot afford to ignore the Challenge of the Stars.

SOCIETIES AND ORGANIZATIONS OF INTEREST TO READERS

The British Astronomical Association (BAA)

Burlington House
Piccadilly
London
W1J 0DU
UK
Tel: 0207 734-4145
www.britastro.org/main/index.html

The British Interplanetary Society (BIS)

27/29 South Lambeth Road
London
SW8 1SZ
UK
Tel: 0207 735-3160
www.bis-spaceflight.com

The International Association of Astronomical Artists (IAAA)

US and World Memberships via:
Kara Szathmary, FIAAA
IAAA President
White Dwarf Studio
119 Marin Drive
Panama City, FL 32405, USA
850-785-4377
membership@iaaa.org

UK and European Membership:
David A. Hardy
99 Southam Road
Hall Green,
Birmingham
B28 0AB
UK
David@astroart.org
www.iaaa.org
(This also contains many links to other artists, organisations and magazines.)

The National Space Society (NSS)

N600 Pennsylvania Avenue S.E.,
Suite 201
Washington, DC 20003
USA
Tel: 202 543-1900
www.nss.org

The Planetary Society

65 North Catalina Avenue
Pasadena, CA 91106-2301
USA
Tel: 626-793-5100
http://planetary.org

UK Chapter:
www.planetary.org.uk

The Society for Popular Astronomy (SPA)

36 Fairway
Keyworth
Nottingham
NG12 5DU
UK
www.popastro.com

Spaceguard UK

The Spaceguard Centre
Llanshay Lane
Knighton
Powys
LD7 1LWC
UK
Tel: 01547 520247
www.spaceguarduk.com

Other Useful Websites

The American Association of Amateur Astronomers
http://www.corvus.com/

American Astronautical Society
http://www.astronautical.org/

The Astronomical Society of Australia (ASA)
www.atnf.csiro.au/asa_www/asa.html

The Astronomical Society of Southern Africa
http://www.saao.ac.za/assa/

The Astronomical Society of the Pacific (ASP)
www.astrosociety.org/

The Mars Society:
www.marssociety.org

The Royal Astronomical Society of Canada
http://www.rasc.ca/

Space.com
www.space.com

Space Frontier Foundation
www.space-frontier.org

Universe Today Directory of Space Societies
www.universetoday.com/html/directory/spacesocieties.html

and

David A. Hardy's AstroArt Website
www.astroart.org

INDEX

Algol 82
Amalthea 46, 47, 47
Antares 80, 81
Ariel 64, 65
Armstrong, Neil 12
Asteroids 38-45, 38, 40-41
    impacts 42, 43

Beta Lyrae 82, 82
Black Holes 90, 91, 94
Bonestell, Chesley 57
Brown Dwarfs 87, 88-89

Callisto 46, 52, 53
Charon 73, 73, 74
Clarke, Sir Arthur C.6, 9
Clusters, galactic 99
    globular 98, 99
Comets 39, 39, 43, 44, 45
Cygnus X-1, 91, 90-91

Deimos 18. 18, 27
Dwarf, white 94, 99
    red 87, 78-79

Enceladus 59
Eros 38, 38
ESA 15
Eta Carinae 92. 93
Europa 46, 46, 52, 54 , 55

Fomalhaut 82, 84-85

Galaxies, colliding 106, 107
Galaxy, jetting 104, 105
Giant, red 99, 100-101
Hale-Bopp 39
Halley's Comet 39

Iapetus 59, 62, 62, 63,
International Space Station, 6, 9,
Io 48, 50-51

Jupiter 46-55, 49, 50-51, 55

Lowell, Percival 19

Mariner probes 20, 32-33, 36
Mars 18-29, 18, 20, 23,
base 21, 21
canals 19
    canyons 23, 23
    exploration 24-25, 26, 27
    polar caps 21, 21, 26, 27
    terraforming 27, 28-29
Mercury 36-37, 36, 37
Milky Way 76-79, 76-77
Mimas 58, 59
Miranda 65, 65, 66-67
Moon 12-17,
    base 12, 13
    exploration 15, 15, 16- 17
    probes 14, 15

Nebulae 102 103
Neptune 68-71, 68
Nereid 69, 69
Neutron stars 94
Novae 94, 94, 95

Omega Centauri 98
Öpik, E.J. 18
Orion Nebula 102, 103

Phobos 18, 27, 28-29
Pluto 72-75, 72, 73, 73, 74-75
Proxima Centauri 78-79, 78-79
Pulsars 94, 96-97

Radio Telescope 12, 108

Saturn 56-63, 56, 57, 58-59, 60-61, 62, 63
Schiaparelli, G.V. 18
SETI 108
SMART-1 14, 15
Smith, Ralph (R.A.) 9
Spaceguard Project 43
Space Stations 8, 9, 9, 10-11,
Supergiant, red 81, 81
    blue 91, 90-91
Supernovae 94

Tau Gruis 86, 87

Titan 57, 57, 59, 60-61
Trifid Nebula 107
Triton 68, 68, 69, 70-71

Umbriel 65
Uranus 64-67, 64, 66-67

V471 Tauri 83

Venus 30-35, 30, 32,. 33, 34-35
greenhouse effect 32
Viking 20-21, 20,
Volcanoes 20, 22, 32, 33, 33, 34-35, 48, 49, 50-51
Von Braun, Wernher, 9
Voyager 48, 56, 59, 64, 65. 68. 69

Zeta Aurigae 81

ARTIST'S ACKNOWLEDGEMENTS

It is almost impossible properly to acknowledge all of the people and influences who have made this book possible. I suppose I should start with Chesley Bonestell, since the discovery of his wonderfully photographic astronomical art around 1950 undoubtedly had a pivotal effect upon my own career. As did discovery of the works of Arthur C. Clarke and those of Patrick Moore himself. But it was Patrick who encouraged me in the early days, by allowing me to illustrate his books and *The Sky at Night*, and by eventually prompting me to branch out into writing my own books. Our major work together was undoubtedly *Challenge of the Stars*, which was of course the starting point for the present book.

Specifically in the preparation of the artwork for *Futures*, I must thank Jo Meder, because the timing of his development of the computer program Terragen for Mac coincided perfectly with my need for a digital 'terrain generator' which I could use in conjunction with Photoshop to speed up the process of producing about 25 of the 80 new illustrations that were required. Jo also kindly lifted the size restrictions on commercial use of 'beta' (testing) versions of this new program. The members of the International Association of Astronomical Artists (IAAA) offered generous and useful criticism (and encouragement) along the way; of those I must single out Ron Miller, Rick Sternbach and astronomer Dr Dirk Terrell.

Finally, a big 'thank you' to my publisher, Cameron Brown, for seeing the potential of this book when others didn't; to Paul Barnett, its editor; to designer Malcolm Couch; and to my wife, Ruth, for not only preparing the index, proofreading and making helpful comments on my own text and captions but for rekeying Patrick's text – originally typed, as always, on his 1908 Woodstock typewriter, but finally transmitted by e-mail. The future has caught up with us!